PERESTROIKA

PERESTROIKA

from MARXISM and BOLSHEVISM to GORBACHEV

Svetozar Stojanović

PROMETHEUS BOOKS
Buffalo, New York

Published 1988 by Prometheus Books
700 East Amherst Street, Buffalo, New York 14215

Library of Congress Cataloging-in-Publication Data

Stojanović, Svetozar.
     Perestroika: From Marxism and Bolshevism to Gorbachev.

     Includes bibliographical references.
     1. Communism—Soviet Union—History. 2. Soviet
Union—Politics and government—20th century.
3. Gorbachev, Mikhail Sergeevich, 1931-   . I. Title.
HX313.5.S76   1988   335.43′0947   88-12620
ISBN 0-87975-488-5

Printed in the United States of America

For
Andjelka, Dusan, and Srdjan

Acknowledgments

This book is the result of my work within the research project "Comparative Investigation of Various Conceptions of Socialism" in the Center for Philosophy and Social Theory at the University of Belgrade, Yugoslavia. I am very grateful to all of my colleagues at the Center, and especially to Zagorka Golubvic who served with me as co-director of the project. Special thanks go to Mihailo Markovic who offered detailed comments and suggestions for improving the manuscript. I also want to recognize and express my appreciation to Eugen Pusic, Andrija Kresic, and Slobodan Inic for their helpful comments.

Of enormous benefit to me was my fellowship at the Wilson Center in Washington, D.C., which extended from September 1985 to June 1986. During my stay at the Center the manuscript was read by many colleagues and friends, all of whom made valuable suggestions, many of which I accepted while preparing the final draft. Unfortunately, I can mention here only a few of those who offered their help: Brian Bennett, Richard Bernstein, David Crocker, Richard T. DeGeorge, Ernest Erber, William Galston, Monika Helwig, Nicholas Lash, Norman Levine, Adam Podgorecki, Jim Satherwhite, Paul Shoup, and Robert C. Tucker.

7

I owe many thanks to William Blackwood, my research assistant at the Wilson Center, and to Steven L. Mitchell of Prometheus Books for his work in preparing the manuscript for publication.

While teaching at Washington University in St. Louis, Missouri, during the period September 1987 to May 1988, my colleagues Carl Wellman, Steven Schwarzschild, and Murray Wax read the manuscript and made useful suggestions for improvements. I am also obliged to Adele Tuchler and Elsie Glickert, who helped me type the final draft.

Finally, I should mention that this manuscript was published in Serbo-Croatian by the Philosophical Society of Serbia (Belgrade) in 1987.

Belgrade   Svetozar Stojanović
Yugoslavia   June 1988

Contents

10   Contents

Introduction

The Crisis and Fragmentation
of Marxism

To say that Marxism is in crisis is, of course, to say nothing new. Crisis has taken hold of not only ideological Marxism but also its intellectual component, and it does not help to say that the crisis rests with Marxists rather than Marxism. Even Karl Marx's own thought is not exempt: to say otherwise would mean helping to uphold one more cult of personality, and surely we have had our fill of them. Here I do not mean only an uncritical attitude toward Marx, that concentrates primarily on the exegesis of his writings, but also the illusion that the reconstruction and revision of his theory could suffice to build a systematic philosophical and social theory relevant to our time.

In my opinion the only Marxism genuinely alive today is both fragmentary and radical-revisionist. What remains of Marx are scattered insights, albeit very profound and stimulating ones. If there is anything strange in the case of Marxism, it is not its crisis but the fact that an old intellectual creation offers a whole range of insights that still stand: on the owner-

13

ship-economic dimension of social classes, on ideologies, and on forms of alienation, to mention just a few.

Those who want to create theories at the contemporary level must incorporate results of various theoretical traditions, including Marxism. Unfortunately, the arrogance of many Marxists still finds them viewing Marxism as a kind of Hegelian "world spirit" that, through its own self-development, absorbs and incorporates all major contributions made by these traditions.

In the final analysis, the very term "Marxist" signals a personality cult. Today disassociation along Marx's lines—"I am not a Marxist"—no longer suffices. A *new* name is needed for those who still derive greater stimulus from the Marxist tradition than from other orientations, even though they tap the latter as well. Unfortunately, no adequate term comes to my mind.

In the history of Marxism, there have been a number of unsuccessful attempts to list the conditions necessary for being considered a Marxist. Fortunately, the most creative Marxists never recognized such an untouchable "essence." In the end, of all the principles the most important remains the search for truth, and with it the principle of *merciless revisionism*

Since Marxist study of the historical process is distinguished by the paradigm of "socio-*economic* formation," it is important to determine the depth of the crisis of Marxism in relation to the basic formations—capitalist and statist—that dominate our world. Constant revision and perfection of this paradigm is of no help—we have to accept that it has limited scope. This means that for each *social* formation in history its specific dominant factor and its precise nature needs to be established. Since capitalism doubtless constitutes a social totality with economic dominance, its corresponding paradigm is socio-*economic* formation.

But history also knows of formations with noneconomic dominance, for instance "socialist" statism, which belongs to the family of socio-*political* formations. Pre-Stalinist, Stalinist, and post-Stalinist statism represents a social totality with political dominance (both in diachronic and in synchronic terms). The examination of the way such statism originated shows that a group first took over political power and subsequently extended it to economic and other spheres of social life. And the synchronic analysis indicates that whenever the ruling communism experienced an antagonistic conflict between political and economic "rationality," it was the former that took the upper hand. In other words, the attempts at more radical economic reforms are ultimately destroyed by the monopoly on political power. It does not mean, of course, that we are dealing with completely immutable states and societies, but rather with very conservative ones. Here political dominance takes the form of *monopoly control over the state and (through it) over the means of production.*

It is easy to reject "socialist" statism in the name of classical humanist Marxism. The more difficult task, however, is to develop a critical theory of statism. But how is this to be done within a tradition that accepts the paradigm of "socio-*economic* formation"?

Critical analysis that points to the *one party* character of the system barely scratches the surface: The question is one of *class* and not just party monopoly. There are still Marxists who entertain the naive hope that statism is a *deformation* in the "transition period" between capitalism and socialism, and not a new class *formation.*

One of the forms of ideological (self-)deception employed by Marxists is the Yugoslav official catchword "socialism as a worldwide process," which is not sufficiently rooted in reality and diverts attention away from statism as a worldwide process. In a large part of the world the working class and the general population have to deal with statism as a socio-political formation, while Marxist ideology continues to mislead them with the model of socio-economic formation. The theory that applies this model, willy-nilly performs an ideological function in statism. Thus, Yugoslav ideologists claim that statism cannot be overcome until the working class "masters economic reproduction." But this is only a secondary lever disguised as a primary lever: the working class is unable to achieve this mastery precisely because politocracy dominates the state and society and, through it, the economy.

"Socialist" statism in many ways is below the historical level achieved by democratic capitalism. It seems that in such statism only liberalization, not democratic socialism, is realistically on the agenda. But, whichever way we look at it, liberalization of statism cannot be a *Marxist* program. What are Marxism's specific ties to liberal ideas, a market economy, decentralization, and the like?

In states that call themselves socialist, Marxists should concentrate on the critique of Marxism as theory, ideology, and practice. They have a real chance to say something new about the society and the spiritual corpus they know from personal experience. The second reason is one of emancipatory interest: Capitalism has never managed to bring Marxism into question to the extent that the existing "socialism" has done. And there must be some kind of ethic of social thought: for what kind of intelligentsia is it that could exist in the statist East, and with state support and recognition concentrate on a critique of the capitalist West?

On a list of contemporary democracies we find, as a rule, the states with the capitalist mode of production. Many Marxists still confront insurmountable ideological obstacles when exposed to real democratic capitalism. But Marxism's theoretical and practical difficulties, stemming from the challenge posed by democratic capitalism, are to a certain extent rooted in the work of its founding father. In his *theoretical* formulations, Marx underestimated capitalism's ability seriously to democratize itself.

One of the hallmarks of living Marxism today, apart from

its fragmentary character, is its radical revisionism. This means that not only are essential statements increasingly subject to revision, but so is the *type of theorizing*. Let me indicate the direction I favor in this regard. As compared to Marx's dialectic, contemporary dialectic should be entirely open; it should be heuristic only and unconfined by progressivist and synthetic frameworks. The critical theory of society and history should be grounded as much as possible in an empirical base. Finally, interest in the utopia of communist society should give way to efforts at developing the theory of democratic and ecological socialism. This implies an incomparably greater interest in an institutional approach rather than the elaboration of communist-humanist ideals and ideas.

While on the subject of democratic socialism, let it be said that this goal will only be achieved if we draw *all* consequences from the fact that Marxism is just one of several important socialist currents. It is also extremely important that we outline the lessons to be learned from the emergence of new social movements, some of which are unquestionably oriented toward democratic socialism. In addition, Marxists must show far more modesty in confronting the development of the postindustrial information society, where the size and importance of the working class will rapidly decline. However, as long as that class does exist, democratic socialism is not possible without its participation and contributions (but not the *domination* of that or any other class).

1

Closed or Open Dialectic?

A Critique of Marx's Dialectic

Karl Marx was a great thinker, full of internal tensions and even contradictions, not only *between* but also *within* the following levels: metatheory, abstractly formulated theory, and theory applied and modified in research. The common denominator of the most important tensions and contradictions, in my view, is the one between closed and open dialectic. It can easily be illustrated by Marx's anthropology, his view of communism, and his historical determinism.[1]

Marx had an excessively optimistic view of human "essence." He left philosophical anthropology some important observations on the *essential potentials* of man. Still, his rather narrow, value-selective notion of essence made it impossible for him to include the opposite, and no less important, set of human potentials. Marx no doubt knew that man is not only a creative, social, and free being, but also a destructive, selfish,

---

1. See chapters 2 and 7 of my book *Between Ideals and Reality* (Oxford: Oxford University Press, 1973).

and unfree being. His *conceptual* apparatus did not, however, enable him to attribute as much importance to the latter set of human potentials as to the former. In Marx the first set falls into the category of human "essence,"[2] whereas the second is included merely in human "existence." Yet historical experience and science show that humankind, precisely in its essence, possesses both of these opposing potentials. Is not its essence manifested historically in ruthless religious, class, and social struggles and wars? Is a consistent dialectic of history possible at all without an internal dialectic of human essence?

Fortunately, Marx the investigator often broke away from the framework that Marx the theorist defined. Thus, for instance, in seeking to illuminate the dispositions and motivations of historical and political actors, Marx's critique of primitive and despotic communism is more important than his depiction of the generic being of humankind (in the depth of the primitive-communist urge for leveling, Marx saw envy at work). Occasionally even Marx abandoned the aforementioned conception of human essence. In the *Theses on Feuerbach,* he wrote (going from one extreme to the other) that human essence is the ensemble of social relations.

Marx undoubtedly is one of the most uncompromising dialecticians in the history of human thought. Yet as he turned from the past and the present and looked toward the future, he did abandon dialectics at times. In his vision of communism, one occasionally senses a tension between the dialectical inclination and a utopia of definitive de-alienation. Marx the dialectician explicitly denied that communism is the end or goal of

---

2. True, in his later writings Marx almost completely gave up terms such as "human essence" and "generic being." He wanted clearly to disassociate himself from any kind of ahistorical essentialism. For my argumentation it is important, however, that he continued to express basically the same content by the term "human nature" (also in *Capital*), still embracing only positive dispositions of men.

history. Occasionally, however, he described communism as a society in which all basic contradictions will have been extinguished, including the contradiction between human essence and existence. But, when the change from a class to a classless society is seen in such absolute terms, the dialectic becomes self-defeating, as it foresees a society in which its basic and necessary principle of contradictions no longer holds.

Marx defended naturalistic determinism, according to which social laws function like "natural laws" with "iron necessity," but also a substantially milder form of determinism in which social laws are only tendencies. In order to distinguish himself from the utopians, Marx attempted to ground his socialism in science. Unfortunately, he felt obligated to demonstrate that the fall of capitalism, and the succession of socialism, was not only a historical possibility and tendency but a "natural necessity." In passing, how could Marx and Engels have spoken of the "naturalness" of socio-historical development and simultaneously of the deterministic dominance of the economic base that has *no exception* ("in the last instance")? Is not the incapacity of some societies, and even whole civilizations, to remove their own "superstructural" obstacles to the development of material production also one of the manifestations of that very same "naturalness"?

As a dialectic of *human practice*, not of *"world spirit"* (Hegel), Marx's dialectic *should* have left open the possibility of either regress or progress. In my view, however, Marx's dialectic is, in essence, Hegelian, to which Feuerbach's transformative method was indeed applied, though not quite consistently and radically. In other words, the *progressivistic* framework of Hegelian dialectic remained unquestioned by Marx.

Aufhebung is one of the central categories both of Hegel's and of Marx's dialectic. As is well known, it means negation, preservation, and elevation to a higher level of development. The *cunning of reason* (*die List der Vernunft*) is a metaphysical guarantee of progress in the Hegelian dialectic. In Hegel, of course, there is no room for the *cunning of unreason* (*die List der Unvernunft*).[3]

A completely open dialectic of *human practice* would have to allow for *Aufhebung* as well as *Anti-Aufhebung*. The latter antonym—one I proposed several years ago—means negation, preservation, and falling to a lower level.

I see dialectic as a kind of heuristic device rather than as a set of empirical generalizations or scientific laws (so-called dialectical materialism). Dialectic does not represent a set of schemes for explanation and prediction. The introduction of a *heuristic* category of *Anti-Aufhebung* into dialectic could stimulate reconsideration of the idea of *internal limitation* (barrier). We should not forget that an internal limitation to further progress may, at the same time, be the internal limitation (barrier) to eventual regress. Such an awareness could prove to be important for the dialectic of human liberation.

These are examples of *Anti-Aufhebung*: First, the negation of the contradiction between the social character of production and the private character of appropriation through "primitive communism." Second, the negation of the separation of the state and "civil society" by the statization of total social life (Stalinistic statism). Third, the "development" of Marxism from Marx through Lenin to Stalin.

Let there be no misunderstanding: of course Marx knew of regress in history. My criticisms have *conceptual* character; namely, Marx's *dialectic as dialectic* can conceptually incorporate only progress. To put it differently, regress is for Marx a dimension of historical process but not of its dialectic. My

---

3. Kurt Wolf, in his article "On the Cunning of Reason in our Time" (*Praxis*, 1-2/1971), rightly reminds that, in addition to the cunning of reason (a force that turns one's bad intentions into good consequences), we have to take into account also the cunning of unreason (a force that turns one's good intentions into bad consequences).

suggestions correspond to the more realistic spirit of our time (in comparison with Marx's time) in which, because of the most tragic historical experiences, important categories have increasingly been supplemented by counter-categories: e.g., negative utopia and negative charisma. After all, a theory in which categories with a negative, critical connotation (such as naturalness of historical process, prehistory, alienation, reification, fetishism, ideology, etc.) play such an important role should be able to accommodate the dialectical antonym.

Unlike the dialectic of the "world spirit," the dialectic of human practice should also be entirely open for *nonsynthesizing* forms of activity and social process. Indeed, the synthesizing categories of *Aufhebung* and *Anti-Aufhebung* can play no central role in such a dialectic. As he ceases to be the (Hegelian) predicate of the self-development of the "world spirit," the man in Marxism admittedly becomes the subject of the historical process; but the point is that the practice of the subject, acting on the principle of *Aufhebung,* is better suited to the "world spirit" than to human beings.

That is why I support those who do not accept the formula of the "materialist *turning over*" of Hegel's dialectic, since we cannot completely separate the dilectic from its content, i.e., the self-movement of the "world spirit" as the material, formal, efficient, and teleological cause of overall development. The dialectic of *Aufhebung* is much more suited to the theoretical constructivist-philosophical domain than to social practice.

Many Marxists still imagine that difficult problems can be resolved by means of *Aufhebung.* It has already been pointed out that Marx himself was sometimes prone to proclaim what he considered to be overcome in the *logical-theoretical* sense was also overcome in the *historical* sense. The consequences of such an approach to democracy in capitalism (even at the time when democracy was in its cradle) are well known.

Even the most revolutionary practice constitutes just one of the forms of *human* practice, which is why we cannot fit revolutionary practice into the logical patterns of the self-movement of "world spirit." Historical actors often opt not for *Aufhebung* of opposites but for preserving them in a new configuration: combination, mutual complementariness, correction, balance, compromise, and control. Even more often the unintended outcome of their activity turns out to be such.

The illusion has to be dispelled that from opposing social entities the positive aspects can be separated at will by means of free de-construction and then built into a new totality. We must not count on the possibility of completely *eliminating* inner limitations of opposite entities; instead we should focus on combining such entities with a view to reducing the negative effects of these limitations to the *inevitable minimums.*

Upon reflection, surely there is no question that all realistic models of democratic socialism must have conflict as a fundamental component of their character. The wisdom in building a new society rests largely in finding the right balance between opposing principles. Here are two examples: first, criticism of the idea that socialism can "dialectically transcend" (*Aufhebung*) the separation of "civil society" and state; and second, rejection of the assumption that socialist democracy can mean "dialectical transcendence" of representative democracy by direct democracy.

The Solidarity movement in Poland is increasingly interpreted as an attempt to establish a *socialist civil society* separate from the state. One characteristic of "civil society" in socialism should be market competition among self-managing enterprises based on social ownership of strategic means of production. Such property, however, should be merely dominant: there should be much more room than there is now for both co-operative and private property. I would call this *socialism with the face of "civil society."*

Instead of the "dialectical transcendence" (*Aufhebung*) of one type of democracy (representative) by means of another (direct), the idea of the "pluralism of democracies" becomes increasingly relevant. Various kinds of democratic participation and representation—of citizens, nations, producers, consumers, and others—should create a specific system of mutual limitations, of checks and balances.

Thus, parliamentary representation, with its "iron law of oligarchy," should be combined with, for instance, a system of councils in which far more direct democracy becomes possible. And, on the other hand, to reduce the effects of what can analoguously be called "the iron law" of manipulation of the unorganized majority by the *de facto* organized minority (in councils), parties would be allowed to organize and work in a visible and formally defined way to gain the majority in the council system.

Those who stand for councils *without parties* should ask themselves why this type of democracy fails as a rule and gives way to one-party dictatorships or, in the best of cases, is reduced to participation at the work place. Anarchism has never managed to show persuasively how councils, without the competition of political parties, can thwart the danger that one group might *monopolize* the state.

MARCUSE'S AND BLOCH'S REVISION OF DIALECTICS

My major concern here is obviously quite different from that of Herbert Marcuse. His main problem, as presented in the essay "The Concept of Negation in the Dialectic,"[4] consists in the following phenomena in "late capitalism": neutralization and suspension of forces of negation, the absence of revolution, and the alleged integration of the working class into the capitalist system. Marcuse embraces them with the formula "stalemate of the dialectic of negativity." Marxism, in his opinion, underestimated the cohesive force of technical advances and science

---

4. *Telos* no. 8 (1971): 130-133.

and their influence in shaping and satisfying needs. For this reason "new, revised dialectical concepts" should be developed that would set out from the *present-day* capitalism.

In connection with this resilience of capitalism, Marcuse analyzes the difficulties posed by the Hegelian origin of Marx's dialectic. Long before Marcuse, many Marxists drew attention to the "positive and conformist character" of Hegel's dialectic, but Marcuse takes this still further, saying that "in the Hegelian dialectic negation takes on a false character. Notwithstanding all the negation and destruction, it is always being-in-itself which ultimately develops and rises to a higher historical level by negation." This is not just a question of adjusting a thinker's views (i.e., Hegel) in light of social circumstances; it is a question of the fundamental feature of his dialectic, in which "the possibility of reason and progress eventually prevails."

Rejecting Louis Althusser's claim that Marx completely broke with Hegel's dialectic, Marcuse says that because of the understanding of the *negation of negation* and *totality,* "even the materialist dialectic remains under the influence of the positivity of idealistic reason so long as it does not destroy the concept of progress whereby the future is always deeply rooted in the present; so long as the Marxian dialectic does not radicalize the concept of transition to a new historical level, i.e. the reversal, the *break* with the past and the present, the qualitative difference built into the theory's tendency for progress."[5]

Marcuse completely shifts the emphasis of dialectical negation from continuity to rupture. And he starts with his perception of the situation in the United States. One must ask how his radical program can be reconciled to his image of "one dimensional" man and society. Under such conditions, Marcuse's hope in a radical dialectical leap cannot be founded.

The revision of dialectic under the influence of life in the

---

5. The italics are mine.

United States is bound to be different from a revision based on experience with "totalitarianism"—both in its Nazi and in its Stalinist form—which completely rejected bourgeois democracy. My concern, like Marcuse's, is motivated by concrete historical reasons. But my point of departure is that the danger of *Anti-Aufhebung* (as I have called it) increases greatly if the emphasis on dialectic is placed simply on breaking with bourgeois society.

Marxists must never forget the tragic consequences that resulted, for instance, from rejecting the New Economic Policy in favor of terrorist "collectivization." We must face the fact that in not a single case of a forced *complete* break with the bourgeois political and cultural tradition was a happy end achieved. Indeed, in the West the Left is pursuing a policy that is the opposite of what Marcuse advocated: now even the eurocommunists have completely shifted the accent from discontinuity to continuity.

While Marcuse's main concern was the *conservative continuity*, Ernst Bloch became increasingly preoccupied at the end of his life with the problem of evil (mainly in the form of nazism and Stalinism). Coming to full expression in the antithesis, evil results in "pure negation," "destruction," "deterioration." His article entitled "Historical Mediation and Novum in Hegel" is devoted to this kind of problematic.[6]

Bloch criticizes the coupling of negation and progress (conceived in a deterministic way) in Hegel's dialectic: "*And why then does negation necessarily find itself on the way to good? And have all negations been creative?* Hegel himself listed as examples of non-progressive (*nicht weitertreibende*) negations the Peloponnesian War and the Thirty Years' War, both of which were as dialectically superfluous as midwives and indeed did

---

6. Published in German as "Geschichtliche Vermittlung und das Novum bei Hegel," *Praxis*, nos. 1–2 (1971).

not allow anything to be born which would be worth mentioning—they were pure destruction. . . . In Hegel's negation there exists primarily a sort of certainty in the combination (*Dreitakt*) of thesis, antithesis, and synthesis. This certainly creates the necessity of the creative, constructive way and a formation of the world which is *guaranteed to head to a good end.*"[7]

Bloch's conclusion, drawn from his criticism, is as follows:

> Not everything is bad, there is much good. There exist in the world wonderful things, and close thereby we find plague, Nazism, and the threat of the Vormärz. Thus, an antagonistic force stalks the earth, and in Hegel's negation it is calmed, made small and friendly. Negation is here only a servant, helping with the breakthrough of good—"a part of that force which constantly desires evil and *constantly* creates good"—constantly must be emphasized. Thus, *the category of evil, the understanding of evil is absent. Philosophically, its absence creates for us even to this day great difficulties.* . . . If there exists a process, a process of atonement, a process of healing, a process of judgement, as a world phenomenon, there must also exist something criminal in it which disturbs us. *And this criminal force, which has been so nicely placed in the middle of thesis and antithesis, has not been comprehended.* To this day we do not have a sufficient concept for it. Why, I ask, is hope not *eo ipso* confidence, but loaded with the category of danger, loaded with the possibility of disappointment, loaded with the possibility of heading to the abyss as mere abstract hope, as wishful thinking?[8]

It is symptomatic that Bloch, one of the greatest philosophers of hope, tried to find a counterbalance to Hegel in Schopenhauer:

---

7. Ibid., p. 18, (emphasis mine).
8. Ibid., p. 19f., (emphasis mine, except for the word "constantly").

> He (Schopenhauer) replaces god with another god, a god of nega-
> tion or negative god, the will to live, the satan through which
> he explains all of the horrible processes in the world. . . . *Although
> this is in another manner again mythologized, at least Schopen-
> hauer takes note of something which in Hegel's negation is beauti-
> fied and covered up by the guaranteed resurrection on Easter
> Sunday.*"[9]

Why is it that Bloch did not reevaluate and revise Marx's
dialectic in the light of his critique of Hegel? It is from his
insight into the enormous power of evil that we have to draw
consequences for our view of man. Is it not true that Stalinism
was also the realization of some of the basic potentials of "human nature"? If we introduce evil into *dialectic of history* as
a force at least on a par with good, we have to make the same
kind of change in our *dialectic of human nature.* Within Marxism
it would mean a radical change not only in *philosophical anthropology,* but even more so in (what I call) the *anthropology
of power.*

Classical liberals start from the worst assumptions about
men in power and then conceive of a series of countermeasures,
such as the division and control of power. They expect those
who hold power to try to enlarge and abuse that power as
much as possible; (a la Lord Acton) power tends to corrupt—
absolute power tends to corrupt absolutely. True, those liberals
who do not move to democratic socialism are unable to propose
efficient measures against the negative influence that the
private-property class has on political power.

Marxists as a rule used to start out from very optimistic
premises because they believed that during the "transition period"
power (and even dictatorship) would be exercised by a whole
social class whose members for the first time in history would
have no particular interests other than universal human interests.

---

9. Ibid., p. 21, (emphasis mine).

However, there is a big difference between the practical consequences of excessive pessimism, on the one hand, and excessive optimism, on the other hand. Pessimistic extremism leads some to undertake superfluous countermeasures against the abuse of power. If we fall prey to excessive optimism, however, the consequences will be more serious: we shall undertake no countermeasures, or, at best, insufficient ones.

IN THE SHADOW OF APOCALYPSE

Anthropologically, ontologically, historically, and theologically speaking, we have been in an absolutely new situation since 1945. If we start counting from that time, we are now in our forty-second year. This condition clearly points up the crisis not only in Marxism and Christianity, but in all ways of thinking, feeling, and acting to date. Even in its most revisionist forms, Marxism seems to have forgotten about man's consciousness of his own mortality and his ability to suppress it. This potential also exists at the collective level: even though we live with the real possibility, and even probability, of collective self-destruction, we go on thinking, feeling and acting as though this absolute *novum* did not exist. Marx's premise about the self-producing being of man is ironically confirmed: we have added collective self-destruction to our vast potential.

In this context, mention should also be made of some *obsolescence of the problematic of human evil*. The accent should be shifted from the topic of evil to the *unconcern, indifference, nonchalance, and carelessness* of humankind. Who suffers from the nightmare of apocalypse? Awareness of its possibility is repressed so much that it cannot manifest itself even in daydreaming. Given these traits, do we have any real chance of avoiding an apocalyptic fate? Confronting us is a new and the most dangerous form of collective self-delusion.

Most important of all, the self-destruction of the human

race can happen quite by chance. The easiest thing, therefore, is to be a fatalist and say that in the final analysis nothing can be done against such a chance event anyway. Humankind can *accidentally* cause *absolute evil*. Does this mean that the Christian satan has chosen contingency as his medium? On the other hand, Marxists take a structural approach to alienation, but it proves useless here: the question is how to deal with the possibility of contingent and yet absolute alienation. Scientists calculate probabilities, but what value or consolation is there in being able to calculate the probability of an apocalypse?

As I have already said, we are living in an entirely new ontological situation. Marx spoke of two kingdoms (realms): the kingdom of *necessity* (material production) and the kingdom of *freedom* that develops on the basis of the former (material abundance). Of course, he did not expect the socio-historical being to remain divided in world terms: in one part of the world there is real hope in the kingdom of freedom, but most of the world still lives in *poverty* in the kingdom of necessity wherein the kingdom of *non-freedom* is rooted.

It is not my intention to dwell on this contradiction in the socio-historical being or on the fact that this may well be one of the causes of humankind's self-reckoning. The point I am trying to make is that a third kingdom should be added to the socio-historical ontology: the kingdom of *contingency*. This is the *framework* of living, acting, and feeling, in which humankind can only hope and try to survive and attain the kingdom of freedom. But what kind of freedom is it that is constantly under the threat of radical contingency? After all, freedom cannot be defined as recognition of contingency. And yet, we ought to add to the *principle of contingency* the *principle of active hope.*

It is the danger of *simple and absolute negation* and not the question of the possibility or impossibility of *dialectical negation*—whether in the form of *Aufhebung* or *Anti-Aufhebung*—that should be of real concern to us. A grain is trampled by man: this example of simple negation, so dear to Engels, appears so naive today. We are now living in the shadow of the possible annihilation of the whole human community. Many Marxists still write about the negative dialectic of human intentions and goals resulting in their opposites. But what about the possibility of an irrevocable and at the same time accidental triumph of negative dialectic over humankind?

The apocalyptic rebellion of things against their human creators may be in the offing. This would be the anthropological version of the Last Judgment. A cynic would say that the final disappearance of alienation can coinicide with its final and absolute triumph. "Dialectical materialists" are left to delude themselves with a dialectic of nature without human beings.

Marxists hope that humankind will manage to overcome the "naturalness" of the historical process, i.e, its blind occurrence "behind man's back." This would be a fundamental turnabout from "prehistory" to the start of real history. We would at last find the key to the enigma of history, in terms of both understanding and mastering it. We are going in the opposite direction, however, toward absolute prehistory: having been at least a kind of subject, we have now become a toy in the hands of contingency for which we have only ourselves to blame. How can we speak of human dignity and of holding our heads up high in the face of radical contingency!

The problematic of utopia also needs to be posited differently. It is ironic but true that at present the most radical utopia right now is the survival of mankind. But is there any more minimal utopia than that! Negative utopia would talk about an earth from which people had eliminated themselves. However, this leaves even science fiction speechless, let alone Marxism and Christianity.

What is the purpose of politics today? Traditionally speaking, politics is concerned with power over individuals, society, and nature. Contemporary politics is confronted with the powerlessness of humanity in the face of chance. The legitimacy of governments, authorities, states, and politics itself appears quite different in this light. A policy that is not preoccupied with the question of human survival cannot be legitimate.

We are also witness to the fact that democracy has lost much of its substance. If democracy is the possibility for citizens to influence the conditions under which they live, then what is its purpose if those citizens influence every other decision *except* a decision about all decisions? Michel Foucault has already criticized the inherited view of power and has pointed out that, in the modern world, power is quite diffuse and spread out, like capillaries. But, I am thinking of the far more radical diffusion of power *and* powerlessness. Where is power *concentrated* that could accidentally destroy humankind, or prevent that very same destruction from happening?

We have to change our approach to justice as well. Some Marxists have been quite critical of those communist regimes that sacrifice the happiness of present-day generations in the name of future generations. In some cases, however, present generations make decisions and permit states of affairs to exist that can prevent not only happiness, but also the emergence of future generations. Would this not be absolute injustice?

The philosophical dispute between existentialism and essentialism is also largely passé. Existentialism came on the scene by criticizing many earlier philosophers for their essentialistic approach: existence precedes the essence of man, not the other way around. However, in our day and age both human existence and essence are open to the whims of contingency that human beings have brought upon themselves. I would call this *transpolitical philosophy contingentialism*. There is the real possibility (even probability) of the circle of human contingency closing once and for all: from human biological contingency in the universe to the contingency of human survival, thanks to humanity's own creativity.

Christianity certainly cannot accept the contingency of humanity's appearance in the universe. If humankind destroys itself, however, it will commit an *absolute, irreparable, unforgivable, and irredeemable sin.* In my view, Christianity is unable to permit such a possibility, which, of course, does not make it any the less real. What if humankind needs a Christianity that would concern itself less with *original* and more with *definitive* sin?

I do not deny, of course, that other possible forms of religion would not stand in contradiction to the threat of the apocalypse. For example, some contend that there was once an Indian tribe whose members believed that God had not finished creating the world and that he would return to it when people had disappeared.

In my opinion, however, no Christian theology, no matter how revisionist it may be, can see the Judgment Day as human judgment in disguise. This self-apocalypse brings into question other Christian teachings as well: divine love, mercy, and salvation, none of which can be unconditional. What theodicy could justify the Christian Almighty? He himself would "die" if the human race destroyed itself! Have we not already committed an unforgivable blasphemy by coming to the brink of the abyss and teetering over it?

It is high time for the whole of Christianity to give absolute priority to a concern for the continuation of humankind. *A theology of the survival of humankind* would be more radical than the *theology of liberation.* Everyone should join the cry of the famous Christian socialist Walter Dirks: "A good end is improbable but imperative."

Christians (especially the statesmen among them) who are nonchalant about the real danger of the human race's self-destruction are no better than irresponsible atheists on the other side of the ideological divide. The world does not need a *Strategic Defense Initiative (SDI)* but rather a *strategic historical initiative (SHI)* for the survival of humankind.

Atheists should give deep thought to whether the notion of a completely *profane* world offers us any chance at all to

avoid the apocalypse. Is it possible to have an atheism for which
the continuation of the human species would be a *sacred* cause?
The arrogant and offensive atheism of "dialectical materialists"
is, in my view, nothing more than a form of what Marx called
"crude and despotic communism." In such states, museums of
"scientific atheism" rest on ignorant attitudes toward religion,
which are perpetuated in schools and other educational institutions. It surely never crossed Marx's mind that his radical
critique of religion—admittedly simplified in places—could become an ideological justification for atheism as a party-state
(quasi)religion.

In the spirit of statist atheism, citizens are educated to *deify*
the state-party and the party-state. The group that intends at
all cost, to hold on to its monopoly in controlling the state
and other key areas of societal life, demands total devotion
to itself and persecutes any rival outlook on the world, including
the religious outlook.

If, as is officially claimed in such states, religion were truly
a completely private matter to which the state is indifferent,
and if the church were truly separate from the state, then
atheism would not be the precondition for climbing the ladder
of social power, and worshippers would cease being second-
class citizens. Who will believe such regimes when they invite
worshippers *outside their control* to cooperate on an *equal foot-
ing* in the struggle for peace and the survival of humankind?

2

Ruling or Dominant Class?

CAPITALISM: A FUNDAMENTALLY
NOVEL TYPE OF CLASS SOCIETY

How can one close the gap between two basic and opposite approaches to society? T. B. Bottomore depicts the conflict as follows:

> Marx's transformation of Hegel's thought involves principally a rejection of the idea of the state as a higher universal in which the contradictions of civil society can be overcome, and an assertion of the dependence of the state precisely upon the contradiction, within the capitalist mode of production, between wealth and poverty, and hence upon the conflict between the two classes—bourgeoisie and proletariat—which embody these contradictory aspects of society. The state is thus conceived as a dependent element of a total social process in which the principal moving forces are those which arise from a particular mode of production. But there is another style of political thought which treats the relation between civil society and the state in a different way, and contributes to a different version of political sociology.

An early expression of this alternative is found in Tocqueville's "new science of politics [which] is needed for a new world," a science which was concerned with the development of democracy, and the formation of a "modern" society (by contrast with the ancient regime) in France, England, and America. The distinctive nature of Tocqueville's conception can be roughly indicated by saying that in observing the two revolutionary currents of the eighteenth century—the democratic revolution and the industrial revolution—which were creating the "new world," he, unlike Marx, paid more attention to the former and attributed to it a greater significance in the shaping of modern societies. For whatever the sources of the democratic movement might be, its consequences, he thought, were clear: its main tendency was to produce social equality, by abolishing hereditary distinctions of rank, and by making all occupations, rewards, and honors accessible to every member of society.

Tocqueville did not ignore the context of industrial capitalism in which the democratic movement existed, as may be seen especially in his analysis of the revolutions of 1848, but he attributed to a democratic political regime . . . an independent effectiveness in determining the general condition of social life. This idea of the autonomy of politics was elaborated by many later thinkers, in a more conscious opposition to Marxism, and came to constitute one of the major poles of political theory from the end of the nineteenth century.[1]

I would like to add that an expression of the complete dependency of politics upon economics within Marxism can be found not only in the model of "base" and "superstructure," but also, I believe, in the application of the concept of the "ruling" class to the bourgeoisie. Marx drew his generalizations about history mainly from an analysis of the transition from feudalism to capitalism. He based his historical paradigm on the analysis of the *capitalist* model of development. However, with the con-

---

1. *Political Sociology* (New York: Harper & Row, 1979), p. 9f.

cept of the "ruling" class, he applied to the bourgeoisie the *pre-capitalist*[2] model of class power. He conceptually mixed up very different forms and modes of class domination. That is why I have suggested that a distinction between "ruling" and "dominant" class should be introduced.[3] Unfortunately, they are still used as synonyms. It is to this proposal that I would like to return now, with what I hope will be somewhat improved formulations and strengthened arguments.

No doubt, Marx did emphasize the depth of the capitalist turning point in history. Thus, for example, in *Grundrisse* he speaks only about three types of formations: pre-capitalist, capitalist, and post-capitalist.[4] My claim is, however, that he diminished the depth of the capitalist turning point by characterizing the bourgeoisie as a "ruling" class, by the way he criticized the thesis of the separation of the state from the "civil society" in capitalism, and, finally, by his insufficiently differentiated categorization of capitalism as belonging to "prehistory" together with the pre-capitalist formations. It is not true, as many Marxists would have it, that the relationships between the bourgeoisie and the proletariat reveal the real and full truth about the prior relationships between all basic classes that preceded them. Capitalism, in my view, represents a *radically new type of class production and domination.*

Marx criticized Hegel's conception of the neutrality of the state vis-à-vis "civil society." Indeed, conceived in a Hegelian

---

2. It was Bolshevik theorists and ideologists, especially Stalinists, who later made an even greater mistake by projecting into the bourgeoisie the way the state and the economy are ruled in their "post-capitalist," statist society.

3. "Marxism and Democracy: The Ruling or the Dominant Class?" *Praxis International* 2 (1981): 160–170.

4. It has turned out, however, that a "post-capitalist" society can have more in common with the pre-capitalist society than with the capitalist society.

way, the thesis of the separation of "civil society" and the state does not hold. On the other hand, the nature of class influence on the state in capitalism does not vindicate its identification through one generic concept ("ruling" class) with other forms of class domination over the state.

The idea of the separation of "civil society" and the state, as I see it, implies two basic truths about the *structural specificity* of capitalism. First, the state is not a domain in which the emergence and reproduction of classes takes place in capitalism. This process unfolds in "civil society"; the capitalist class structure has economic and not political character. Second, the most powerful class in capitalism is separated from the state in the sense that it does not have a structural monopoly over it. The bourgeoisie is no state-class. The differences between liberal and organized capitalism have been the subject of much discussion and a significant body of writing. It is true that in capitalism there is an increasing intertwining of politics and economics. However, with regard to the above mentioned features of capitalism, nothing fundamental has changed.

T. B. Bottomore rightly emphasizes that the capitalist transformation is almost without parallel in the whole of history: "But that transition, in the sharpness of its break with the past, its revolutionary significance for the future, has perhaps no historical parallel, unless it be the Neolithic revolution in the early stages of human history. No other historical transformation has quite the same clear-cut and definite character."[5] However, even he does not raise doubts about the Marxist conceptualization of the bourgeoisie as a "ruling" class.

True, capitalism belongs to "prehistory" in that the socio-historical process therein largely takes place without the conscious control of people. However, if one of the lines of demarcation between "prehistory" and "history" would consist in the existence or absence of *class structural monopoly over the state*,

---

5. *Political Sociology*, p. 89.

then we could say that capitalism does not belong to the "prehistoric" formations.

In claiming that capitalism provides the most convenient starting point for the emergence of socialism, many Marxists are referring primarily to the level of material development. It is, in my opinion, even more important that capitalism's separation of "civil society" and the state represent one of the *necessary preconditions* for socialism. Statism is no *Aufhebung*, but rather the *Anti-Aufhebung* (as I defined it in chapter 1) of that separation. As a consequence, within statism the struggle for democratic socialism has to begin at a lower level. In this context it is easy to understand why the view that there is a need to establish "socialist civil society" and separate it from the state is becoming more attractive.

RULING OR DOMINANT CLASS?

We are still living in societies in which some social groups exercise *control* over the means of production. As a result, other groups have to work and produce for them. It is here that the basis of social classes is still to be sought. Marx was mistaken, however, in believing that control over the means of production ultimately rests with those who own property, though this is certainly true for capitalism. There have been cases in which class divisions had a *politico*-economic base: the states of the so-called Asiatic mode of production (and Oriental despotism) as well as contemporary statist ("socialist") societies are representative examples.

Marxism still has a lot to say about the economic-property dimension of class grouping. And yet, an acceptable view of classes has to be much broader. Social classes should be defined by their role in the *mode of production as well as the mode of domination through the state.* Their power or powerlessness stems from their relation to the means of production and the means of exercising the might of the state. In coming closer to the tradition that emphasizes the autonomy of politics I am, however, still unwilling to accept the idea of *purely political* classes.

In my view, the general concept should be that of a dominant, rather than a ruling class. It goes without saying that every instance of class rule is at the same time an instance of class domination; however, not every instance of class domination is at the same time an instance of class rule. Consequently, a class that has to rule in order to dominate should be distinguuished from a class that can dominate without ruling.

In my conceptual apparatus, class "rule" represents an extreme form and mode of class domination. A "ruling" class *directly and exclusively runs the state and through it exerts a monopoly control over the means of production.* Examples of ruling classes in this strict sense would include state rulers and administrators in the so-called Asiatic mode of production (and Oriental despotism)[6] and the contemporary statist class.[7]

---

6. The land and the systems of irrigation were under the control of a despotic state, which means under the control of the class of state rulers and administrators who participated in the distribution of the surplus product in a manner proportionate to their place in a fully centralized hierarchy, usually with a monarch at the head. The peasants lived and worked in mutually isolated communities, and they paid taxes to the state in the form of agricultural products and labor power for public works (primarily for the construction of irrigation projects).

It is clear that the geographical denomination for this mode of production (Asiatic) and this despotism (Oriental) cannot be upheld, above all because researchers have discovered it on other continents as well. My suggestion for a new name is *agricultural-statist mode of production* or *agrarian statism (and despotism)*. The adjective "agricultural" is indispensable if we are not to confuse it with *industrialized* and *industrializing statism*, the archetype of which is represented by Stalinism in the Soviet Union.

7. I have written about this class several times since 1967, but in earlier discussions I mistakenly claimed that it is a collective *owner* of the means of production.

The bourgeoisie, however, is an economic and not a politico-economic class, and thus can be only a *nonruling dominant* class. In this respect there is no difference between democratic and dictatorial capitalism—there is no ruling class in the latter either. Due to the separation of "civil society" and the state, economic criteria do suffice to identify classes in capitalism.

The bourgeoisie seems to be the first class in history that can dominate the state without running and monopolizing it. In this regard as well, capitalism is an *essentially new type* of class society. Indeed, ownership of the means of production provides a great advantage in the class struggle for domination over the state, but this ownership is far from being sufficient to establish and maintain a structural monopoly over the state.

If we want to determine who belongs to the so-called "nomenklatura"[8] in the USSR, we have to first apply political criteria. I am not suggesting, of course, that the "nomenklatura" lacks economic power or that it is politically omnipotent. However, its political power is primary and its economic power is derived from and subordinate to the former. For this reason I maintain that it is a *political*-economic class, unlike the bourgeoisie as an economic class (no political criterion is needed to determine who belongs to the bourgeoisie).

Even in democratic capitalism (let alone its dictatorial form) the state has increasingly been intervening in social life. In introducing the concept of statism, I did not simply mean state intervention, but rather a new social system in which one group enjoys a structural monopoly control[9] over the state and the

---

8. The name itself suggests that this group makes lists of its members and changes their composition by (self-) co-optation and exclusion. A question to be examined and answered separately would be whether these lists are completely identical with the group that (according to *objective* criteria as well) directly runs the state and through it exerts a monopoly control over the means of production.

9. Not every monopoly control, naturally, has to be or become "totalitarian."

means of production. Statization *within* capitalism should not be confused with statism as a *new type* of class system. When a democratic state takes over the means of production, even in capitalism, the takeover is a form and mode (albeit the lowest) of their socialization. On the other hand, when a state-class establishes a structural monopoly control over the means of production, it is no step in the direction of their socialization.

When I introduced "statism" as a new social-political formation, I did not have in mind only its communist variant, although I have been analyzing only this kind of statism. Populations throughout the world are witnessing the emergence of various kinds of statism that have no communist origin whatsoever—for instance, militaristic statism.

There are some nuances in Marx's usage of the notion of a ruling class: he speaks, for instance, about direct and indirect class rule, and even about economic and political ruling classes. All such distinctions, however, do not provide enough room to introduce the concept of *political*-economic class. This concept doubtless goes beyond the framework of Marxism.

It would certainly be possible to agree with the substantive side of my distinction and yet choose a different terminological strategy: for example, to keep "ruling" class as a generic concept and, at the same time, introduce different designations for different kinds of class rule. In my opinion, however, this would not be good, exactly because the difference I have in mind is one of principle, rather than simply of degree. That is why it is much better to make more radical changes in our conceptual framework. I shall discuss in chapter 3 the adverse consequences of the category of "ruling" class as applied to the proletariat in Marx's idea of the "dictatorship of the proletariat" during the "transition period."

My conceptual innovation is not primarily motivated by considerations of terminology, although Tocqueville had a point when he wrote that the human spirit discovers *things* easier than *words*. My proposal is still less the result of terminological hairsplitting; rather it results from the desire to meet important theoretical and practical needs. If adopted, its consequences will be of equal theoretical and practical importance. What, then, are the advantages of distinguishing "dominant" from "ruling" class? In other words, what problems might be more easily identified and resolved when such a distinction is employed?

The concept of "ruling" class is not suitable enough to express the real character of bourgeois power, and it is even less suitable to the recognition of democracy as a reality in capitalism. Unlike "ruling" class, the notion of "dominant" class is sufficiently flexible to avoid such faults; at the same time it is "sharp" enough to suggest that democracy in capitalism has class limitations (I).[10]

It is only within such a conceptual framework that one can fully understand and explain how all men and women have become citizens with equal rights. I agree with Eugen Pusic when he writes:

> Bourgeois-democratic revolutions proclaim in this domain the opposite principle of *inclusion*. The traditional concept of subject, designating those who are politically interested but at the same time are in principle excluded from the legitimate realization of their interests vis-à-vis the authority, the political sovereign, is replaced by the concept of citizen. In principle all people play the role of citizen. All citizens can legitimately and freely realize their political interest under the conditions of fully equal rights. *The institution of citizen differs from the institution of subject not only quantitatively, but rather as a whole it belongs to quite a different universe of institutions.*[11]

---

10. To make an overview easier, basic answers to the above raised question are designated by corresponding Roman numerals.

11. "Self-government and Law," study group "Man and System," vol. 6, The Institute for Social Research, Zagreb University, Croato-Serbian. (The last sentence reflects my emphasis.)

The lack of a ruling class constitutes a necessary, albeit an insufficient, condition for democracy for all citizens. Had the bourgeoisie been capable of structurally monopolizing the state, there would not have been democracy for all citizens. In such a case, the idea of division and control of power would be an illusion because such division and control would remain within the scope of the ruling class.

The historical shift from feudalism to capitalism is characterized by two revolutions: the industrial revolution and the bourgeois-democratic revolution. They pave the way for the elimination of the estates, the transformation of the populace into free citizens, and the painstaking development of democracy. *In his theory,* Marx did not evaluate these two revolutions with equal farsightedness; he conceptually underestimated the *democratic developmental potential* of the bourgeois revolution and the state.

We should not, however, overlook the fact that in Marx's and Engels's time, democracy was just in its infancy. In the mid-nineteenth century the working class was indeed deprived of its rights. Many struggles, often bloody ones, had to be waged before property and other barriers standing in the way of attaining political rights could be eliminated, or before workers could achieve the legal right to strike and to organize unions and political parties, or even before women became full-fledged citizens.

Jean Ellenstien calculated that from the beginning of the French Revolution it took 80 years before universal suffrage was achieved, 95 years before the freedom of the press was instituted, and 112 years before the right of association was acknowledged. For these reasons, in the past century, as well as at the beginning of the twentieth century, it could easily seem quite appropriate to talk of the bourgeoisie simply as one of the ruling classes in history, and to characterize democracy as a form of covert rule or even dictatorship of the bourgeoisie.

Leninist and, indeed, all Bolshevik political culture fully

rejects bourgeois democracy. In Stalinism this became the *sine qua non* of the ideological legitimation of statist dictatorship. Lenin further simplified some of the claims of classical Marxism, such as the claim that the government is the executive committee for carrying out the joint affairs of the bourgeoisie as a whole.

In *ideologized* Marxism, Marx's notion of the *rule* of one class is totally confused with the *dictatorship* of that class—indeed, Marx himself occasionally equated them. For this reason, many Marxists drew the conclusion that every rule of the capitalists is *more or less* the dictatorship of the capitalists. The Leninists then shifted the accent from "less" to "more," and the Stalinists ultimately eliminated any qualification. It is no wonder that communists did not distinguish between dictatorial and democratic capitalism. During the 1930s this fact had catastrophic consequences for the entire world. Thus, in Germany communists used to concentrate on social democrats as their main "enemies."[12] Having found themselves together with social democrats in Hitler's concentration camps, they learned about the enormous difference between dictatorship and democracy—but it was already too late.

Being ideologically blinded, all communist parties had for a long time a completely distorted picture of existing capitalism. Their failure in the West is quite understandable, since they

---

12. This is how it was described by Karl Popper: "After all, since the revolution was bound to come, fascism could only be one of the means of bringing it about. . . . Only the vain hopes created by democracy were holding it back in the more advanced countries. Thus the destruction of democracy through the fascists could only promote the revolution by achieving the ultimate disillusionment of the workers in regard to democratic methods." (*The Open Society and Its Enemies* London, 1969, vol. II, pp. 164f.) What else could be expected from those whose efforts were not concentrated toward the reduction and elimination of bourgeois domination in the democratic state, but rather against democracy itself?

did not have—and some of them still do not have—a clear picture of the real way a democratic capitalist state functions and the real bourgeois impact on it.

The concept of the ruling class is less suited for allowing various degrees of power than is the concept of the dominant class (II). The domination of one class can quite readily be said to be greater or lesser, both in comparison to other classes and in relation to itself on a temporal continuum. Through the struggle for the democratization of the state and society in capitalism, bourgeois domination has diminished, and the influence of other classes, including the working class, has been considerably strengthened.

By suggesting homogeneity, the concept of class rule does not provide enough flexibility for the asymmetry of diverse forms and means of domination: political, economic, ideological, cultural, and so on (III). The bourgeoisie is, however, a class whose influence varies greatly among its subgroups and from one area of social life to another. This holds true especially for the bourgeoisie's ideological impact, which in some areas (in the economy, for instance) can be dominant, and in others it can be small indeed (for example, in high culture). If the bourgeoisie is categorized together with a state-class ("nomen-klatura") as a "ruling" class, the enormous historical, structural, and functional differences between the ideologies of these two classes will not be comprehensible (IV). Chapter 4 will address these differences.

Among other difficulties in Marxist theory is the under-standing of the precise relationship between the bourgeoisie and the power elites: political, business, administrative, military, and the like. Does it make any sense to classify these elites as belonging to the bourgeoisie if they come from other classes and are not owners of capital? Is it really the case that they always represent the bourgeoisie and its interests? What should be done about military dictatorships in capitalism? Put more broadly, what is the relation between military rulers and the "ruling" class? It is also very well known that those who espouse the theory of the bourgeoisie as a "ruling" class have great difficulties examining and explaining fascism and nazism.

It is because of these and similar problems that many Marxists do not use the notion of elite, although it has no doubt proven to be fruitful in sociological, historical, economic, and political studies. Understanding the bourgeoisie as a nonruling dominant class should fully open Marxism to the study of the power of various elites (V). The concept of an elite is very suitable for the analysis of not only bourgeois democracy but also revolutionary and post-revolutionary "vanguardism." It is interesting to note that the ideologues of the Yugoslav Communist party proudly refer to the "vanguard character" of their party and at the same time reject "elitism" in principle, as if the vanguardism of a party, particularly one with a monopoly of power, were not a kind of elitism.

The thesis of bourgeois "rule" focuses too much attention on coercion. The theory of bourgeois "domination" is much more open to the examination of various forms, degrees, and modes of socialization, consent, legitimation, cultural reproduction of social systems, and similar concepts (VI). The Italian Marxist Antonio Gramsci tried to eliminate this shortcoming while maintaining the conceputalization of the bourgeoisie as the "ruling" class. He introduced into his analysis the concept of "hegemony," by which a class, through ideology and culture, obtains the consent of other classes. Thus, one class can preserve complete control over the means of coercion, and yet lose its hegemony. Gramsci no doubt made a great contribution to the development of Marxist thought on the capitalist state. On the other hand, there is the unsolved problem of how to conceive the possibility of significant democratization of the capitalist state if the bourgeoisie is presented as the "ruling" class in it. For Gramsci, bourgeois democracy ultimately remains a form of financial and oligarchic rule, and every state, in the final analysis, remains a dictatorship of the ruling class.

Partly due to Hegel's influence, Marx was inclined to under-estimate the force of morals. He used to say that morals are "incapacity in action." No wonder, then, that dogmatic Marxists paid no attention to the moral legitimation and the moral impact of bourgeois democracy, both of which can best be seen if we compare the moral foundations of bourgeois democracy with the alternative, i.e., violence as a means of social stabilization or change. It is not true, as many Marxists would have it, that common citizens are unaware of the shortcomings of democracy in the West. Rather, it is violence as an alternative to voting that appears unacceptable to them in every respect, including the moral. One revisionist Marxist who has drawn radical consequences from this fact is Albrecht Wellmer:

> Reasons *against* the use of physical force against people, and especially reasons against destruction of human life, have *a priori* character. These reasons have, as a matter of course, to be accepted by every socialist. Consequently, the greater and more irrevocable physical force is, the more it requires legitimation in every and any instance . . . . On the other hand, violence (including destruction of human life) can be morally justified; for instance, in situations in which violence is the only available means to successfully resist unbearable repression and unbearable cor-ruption. In certain situations there is perhaps even a moral obligation of physical resistance or use of physical force. But, he who decides to use force takes upon himself the great burden of justification.[13]

If bourgeois democracy is construed as a form of disguised "rule" of the bourgeoisie, how does one justify the presence of hope in the democratic road from capitalism to socialism? Marx listed Great Britain, Holland, and the United States as exceptional countries in which there existed the chance for a

---

13. "Terrorism and Social Critique," Serbo-Croatian translation from German manuscript, Knjizevna, rec., lo. XII. 1983, Beograd.

peaceful and democratic transition to the new society; Engels added France and Germany. However, all that remained incongruent with Marx's theoretical conception of the capitalist state.

So-called "eurocommunism" has opted for the long march toward socialism through democratic bourgeois institutions. However, a more profound and consistent elaboration of democratic communist strategy demands a radical change in the *conceptual* approach to capitalism. This has not been accomplished even by Umberto Cerroni, one of the principal theoreticians of contemporary Italian communism. In *Political Theory and Socialism* he offers an excellent analysis of two contradictions:

> . . . the contradiction between private ownership (capital) and wage labour, and the contradiction between sovereignty which is delegated to the political *elite* and the formal recognition of the equal political capabilities of all. The analysis of the first contradiction is well known: Marx dedicated the greater part of his life to it. But the analysis of the second remains as yet merely indicated.[14]

Consequently, Cerroni asks for a theoretical analysis "capable of pointing out the ambiguity of democracy, *its potential anti-bourgeois and anti-capitalist character,* thereby strengthening the importance of it for the political system of socialism itself."[15] However, the precondition for such an analysis, in my view, would consist in drawing a conceptual distinction between bourgeois domination and other types of class domination.

Marx expected the working class to overcome the bourgeoisie and become the new "ruling" class. This view has been disputed by critics who begin with a different set of assumptions; namely, they remind us of the fact that the most

---

14. "Political Theory and Socialism," Croat-Serbian translation from Italian, Zagreb, 1976, p. 63.

15. Ibid., (my emphasis).

oppressed classes have never become ruling classes: neither slaves, nor serfs, nor peasants (after the system I call agrarian statism ceased to exist).

I, too, doubt that workers will emerge as a "ruling" class, but my doubts are based primarily on an analysis of the *type of power* the working class possesses. Being an economic (and not a political-economic) class, workers, in my opinion, cannot become a *ruling* class. Had Marx operated with the paradigm of the dominant class, he would have been more prone to raise the question about the basis of eventual working-class *domination* (VII).

Since the working class is not a class whose members own the means of production, its domination would have to rest on special institutional and organizational arrangements. The crucial question is whether the working class is capable of creating organizations and institutions that are sufficiently powerful to prevail over the tendency of the new state apparatus to become independent and even constitute itself into a new *ruling class*.[16] How else can we soberly determine the chances that a particular social group will become dominant if not by comparing them with the means available to the competing groups? Unfortunately, Marx was neither inclined nor ready to raise such questions.

I am not referring here to the usual threats to citizens created by state power as such, but to the potential and the tendency of state power to grow into a *state class*.[17] Thus, it is quite

---

16. Directing attention in the Bolshevik party to the vast majority of peasants in Russia as the main handicap for socialism objectively diverted attention away from the new party-state apparatus as the main threat to the working class.

17. It goes without saying that this class is not identical to the *state-party*. Thus, an analysis and critique, which finds the main problem to be the *one-party* character of the system is, in my opinion, superficial.

understandable that none of those who adhered to the naive conception of the "transition period" in classical Marxism were able to anticipate anything remotely resembling Stalinism.

Marxists are usually reproached for not having done institutional, but only class analysis. The question, it seems to me, is whether one can talk about any *adequate class analysis* in this case if it does not also embrace an *institutional analysis*. Is it really possible without institutional analysis seriously to examine the chance for the domination of an economic class such as workers? In connection with this I would like to voice my doubt that the protagonists of *socialism of councils without political parties* are able to indicate how socialist society could establish domination over the state.

Let me close my list by emphasizing one further advantage of a theory that would distinguish the dominant from the ruling class. This advantage, in a certain sense, sums up all those previously enumerated. There have been many publications and discussions that address the question of why Marxist political theory finds itself in such an unenviable position. Ralph Milliband, in his book *Marxism and Politics*, suggests several such reasons. I would like to add one more: If the state, as the central political institution of the capitalist society, is an instrument of the *ruling* class, then basic political questions are closed before they are even seriously posed (VIII). The same holds true for the assumption that the working class will replace the bourgeoisie as the *ruling* class.

THE INFERIORITY OF STATISM

It is my contention that *an objective assessment of democracy in the West represents one of the most reliable criteria for recognizing the democratic character of conceptions of socialism.* My conceptual revision allows for the appreciation of democracy in developed capitalist societies and, at the same time, for the democratic-socialist critique of the ideologization of such democracy.

Democracy is a powerful source for the legitimization of capitalism, even stronger, perhaps, than high living standards. Democracy also constitutes an important achievement for our whole civilization, a step forward that noncapitalist classes have also put to good use.[18] It enables important social reforms. And finally, democracy defends the capitalist system against radical and abrupt changes.

Contrary to what many dogmatic Marxists think, the bourgeoisie is not compelled to instigate a military *putsch* every time it feels threatened. Its domination is more subtle, negative, and indirect: capitalists are capable of staging, for instance, so-called investment strikes and transferring capital abroad, or inventing any number of ingenius strategies to remain in power. A government that practices anti-capitalist measures will collapse for economic reasons—if not immediately, then at the next elections. Only if such a government tries to violate the principle of elections will the repressive apparatus intervene, but then only in the name of democracy, not in the name of the bourgeoisie.

Private ownership of the means of production as the basis of bourgeois domination enjoys international support in the world market. Barriers pertaining to ownership, education, and sex have long since been eliminated and universal suffrage realized. In the hands of the bourgeoisie, however, the international market has remained a crucial factor in the "capitalist encirclement" of all attempts to introduce statist or socialist changes of the capitalist system. To make a point in passing: statism is much weaker in the world market than is capitalism.

I would like to quote two opposite opinions of democracy.

---

18. In recent times the bourgeoisie has undertaken an offensive against the "welfare state" to regain some lost aspects of domination.

First, the opinion of the president of the Federal Republic of Germany, Richard von Weizsäcker, and the second that of sociologist Claus Offe. For both, the occasion to voice their opinions was the Flick and Barzel corruption affairs. It will be immediately clear that von Weizsäcker is giving an ideological appraisal, whereas Offe's is a critical appraisal. I share Offe's judgment, the more so because he is warning of the danger of the Left criticism (as a result of such affairs) reversing to dogmatic (over)simplifications about democracy in the West.

Perhaps there are representatives of firms who believe that they are able to buy politicians and politics. . . . It is true that the leading politicians in the parties and the government hold talks disproportionately often with members of the boards of big firms. But to draw from this the conclusion that those businessmen have a disproportionate amount of influence is wrong. And I say this knowing what the politicians depend upon in order to gain and maintain their mandates.[19]

With regard to the economic influence on state politics, exceptions to rules merely prevent us from seeing the legal normality of these relationships. The politics of the government are in *its* own interest (among others, financial and economic), even without bonuses in the form of thick brown envelopes, instinctively directed toward guarding, nourishing and favoring industrial interests. Whether *this* normality is justifiable or not concerns very few people in times of scandals and because of scandals. . .-. Power is the capacity to bestow gifts. The voluntary character of gifts is based on the assumption that the gift might not be given. And the fear that the next time the gift to the party will perhaps be withheld functions within the parties constantly as a guiding principle for acting and not acting. Every party is subjected to this normal process of creating dependence; indeed,

---

19. Richard von Weizsäcker, Interview, *Die Zeit*, vol. 11, no. 2 (1984).

every party has a good conscience about its dependence simply because it is done by all parties. With regard to one's own financially intimate sphere, one can count on being spared from exposure by competitors.[20]

Democracy in the West established its principal institutional, procedural, and even cultural contours quite some time ago. Because of structural limitations of a capitalist character, it cannot develop much beyond that plateau. There is already too much power concentrated in political and other elites—in other words, the power of the social base is small. And there are entire spheres of social life (first and foremost the economy) that, due to the structure of ownership, cannot experience genuine democratization. New social movements criticize those who would abandon the idea of government of the people and by the people, and in so doing reduce democracy to democratic elitism.

The desubstantialization of democracy reaches its peak when political elites can make decisions of *fateful* importance regardless of the popular mood. Opinion polls showed, for instance, that 70 percent of West German citizens were against the stationing of U. S. Pershing missiles on their soil, but that did not prevent the West German Parliament from voting to deploy them anyway. But what is the meaning of democracy if the people can have a say on resolving all other questions except the biggest question of all: the question of their own survival and that of all humankind?

One should deduce the need to socialize the *strategic* means of production from the imperative of radical democratization. This means viewing socialism first of all as radical democracy. The primary principle should be this sort of democracy, and the derivative principle should be socalization of the means

---

20. Claus Offe, "Von der Suchtbildlung der Porteieu," Die Zeit, vol. 12, no. 7 (1984).

of production. Anyone who thinks and acts to the contrary cannot be acceptable to either new social movements or, in a more general sense, to democratic forces in the West.

Yugoslav official ideology has not given up a provincial political conceit in the name of "self-government socialism": democracy does, admittedly, exist in the West, but it is ostensibly something entirely different and inferior to self-government as an idea and reality. Fully developed self-government can, however, be nothing else but radical democracy.

The world today is dominated by two systems with gigantic concentrations of power: capitalism and statism. The latter is not only de facto inferior (except in the military sense), but it also feels inferior deep inside. That is why, when confronted with democratic capitalism, it has developed a whole system of ideological rationalizations. This, in my opinion, is a "sour grapes" ideology.

Suffice it to recall the slogan "to catch up and surpass capitalism." However, statists have not been able to catch up with developed capitalism. Since democracy represents a *historical a priori*, they do not dare abandon it openly. Their inferiority complex is manifested in various ways: among them, the statist tendency to both attack and ape democracy in the West. Hence their democratic forms lack substance: fixed nominations, election campaigns the results of which are known in advance, election "contests" with but a single candidate, and so on.

No *ruling* class is capable of governing democratically since this would mean that, by voting, the people could replace it and thereby eliminate it as a class. Statists who, let us suppose, had lost monopoly ruling status in politics, would not be in the position (like the bourgeoisie) to dominate politically through the economy, because in the statist system there is no real division between politics and the economy, and because the ruling economic power of statists is derived from politics. To lose elections would mean the loss of monopoly control over the state and over the means of production.

The argument that democracy is allegedly impossible in statism because of economic and other sorts of underdevelopment cannot be accepted. After all, democracy came into being in capitalism when it was incomparably less developed than many statist countries today. The real explanation is that democracy would threaten the statist *system* as such. That is why the attacks against democratic tendencies are inherent in statism. This does not mean, of course, that "socialist" statism has in every respect to be judged negatively. Statism may perform a progressive function in some countries, for instance that of concentrating accumulation of capital needed for modernization. Some underdeveloped societies really cannot afford the slow process of capitalist accumulation and modernization.

Would it be true to evaluate statism, as some would have it, as a "dead end of history." In "socialist" statism in Europe (except Albania) industrialization and modernization have been successfully carried out, and thus we cannot speak of a "dead end" in that sense. Of course, the price paid for this *type* of industrialization and modernization has to be assessed, primarily taking into account human sacrifices ("pyramids of sacrifice," as Peter Berger would say).

The basic question to be asked now is: Having completed industrialization and modernization, has the statist system become the impediment to further economic and social development? Can it be significantly reformed? One thing is fairly clear: statism is not as dynamic as democratic capitalism. The latter is more quickly and successfully heading toward the postindustrial, information age.

3

Marx and the Bolshevization
of Marxism

RESPONSIBILITY FOR THE MODE
OF THEORY CCONSTRUCTION

In this age of pervasive commercialization, there can hardly
be anyone who has not heard of the younger French theoreticians who, in Nietzschean style, call themselves the "new philosophers." Initially they were ideologized Marxists, but, shocked
by Solzhenitsyn's revelations of the "Gulag," they were transformed into ideologized anti-Marxists. To paraphrase two of
them, A. Glucksmann and B. H. Lévy, we must not ascribe
concentration camps to bureaucracy alone; this "barbarism with
a human face" would not be possible without Marxism. To be
sure, Marxism does not automatically lead to the Gulag; it
"merely" makes people blind and deaf to the emergence of these
camps. The Gulag is the true realization of Marxism, and Stalinism is socialism "transformed to the point of recognizability."

In diametric opposition is the view held by those who proceed from Marx's humanism. This orientation, with the accent

on "returning to the authentic Marx," played a tremendous role in the showdown with Stalinism in the fifties and sixties. It says that Marxism has not been realized; rather, Marxism has been "distorted to the point of unrecognizability" in Leninism, and even more so in Stalinism. Marx bears no responsibility whatsoever for the ideologization of Marxism.

It is here that my reservations begin concerning such Marxists. In my view, Marxists ought to be radically critical toward Marx himself because of his part in the bolshevization of Marxism—critical, of course, in an entirely different way than the "new philosophers." We no longer have the right to judge Marx in utter isolation from the many miscarried attempts to achieve his ideas, no matter how unacceptable these attempts may be to humanistic Marxists.

I still share with these Marxists the belief that bolshevism constitutes an outlook on the world and a practice that has moved far away from the humanistic position of original Marxism. But, it is much easier to reject bolshevism in the name of such Marxism than to explain its bolshevization in the first place.

In the way in which he constructed his theory, Marx, in my opinion, not only opened up the possibility for its bolshevization through reinterpretation, reduction, and transformation, but, to a certain extent, he lent a helping hand. Until recently there were many Marxists who believed in the existence of an essentially homogeneous Marx. However, at its very foundations the edifice of his thought is fraught with tensions, and even contradictions. Originally, Marxism left behind large areas of ambiguity and vagueness, as in the case of the so-called "transition period." Still worse, Marx was responsible for creating an essentially unproblematized and fallacious idea, which, because of its fertile ground for ideologization, was extremely dangerous from the start: i.e., "the dictatorship of the proletariat" during the "transition period."

It should be remembered that Marx designated this idea as his only original contribution to the theory of class struggle.[1] Admittedly, many see an extenuating circumstance in the fact that he explicitly mentioned this idea only about a dozen times, but, in view of the enormous importance Marx attached to it, this is more of an aggravating than an extenuating circumstance.

Marx's ambivalence has often been analyzed and criticized: on the one hand, he advocates self-organization by the proletariat, citing the Paris Commune as a model, but on the other, he talks about communists as the most farsighted part of the proletarian movement. However, without the mediation of the "dictatorship of the proletariat," this second aspect certainly could not in itself be a sufficient basis for the bolshevization of Marxism.

As for (co-)responsibility for the bolshevization, it is entirely natural that in the case of Marx we should concentrate on theory: he was for the most part a theoretician, and it was this theoretical aspect of Marx's work that influenced the bolsheviks. In discussing Lenin's (co-)responsibility (for the *bolshevization of Marxism and the Stalinization of bolshevism*), the accent would quite certainly shift to his practical and political activity, because he was, above all, a revolutionary and the founder of a new state.

Since a disproportionately large number of philosophers have always taken part in discussions on the continuity or discontinuity between classical Marxism and bolshevism, too much stress has been placed on the *philosophical* aspects of the problems: e.g., dialectics, determinism, and anthropology. The uninitiated, therefore, might get the impression that Lenin and his followers were mostly preoccupied with the realization of Marx's *philosophy*.

---

1. Ralph Millibrand rightly points out: "This is a crucial concept in Marx's political thought, and also in Leninism." (T. B. Bottomore, ed., *A Dictionary of Marxist Thought*, Basil Blackwell, 1983, p. 129).

However, the decisive theoretical inspiration for bolshe-vism should be sought in Marx's *socio-political* ideas. That is why the kingpin in my analysis is the "dictatorship of the proletariat," as the "weakest link" in classical Marxism. In renouncing precisely this project, eurocommunists must have perceived the principal nexus between bolshevism and classical Marxism.

"Dictatorship of the proletariat" was the driving idea of the "negative dialectic of Marxism." This, of course, is not to say that Marx's ideas were more decisive for the emergence and victory of bolshevism than the socio-historical conditions prevailing in Russia and the world at the time. Still, because of Marx's enormous authority in social democracy, an important role was played by the ability of Lenin and other bolsheviks to cite the "dictatorship of the proletariat." Certainly, a theory that deliberately takes upon itself the responsibility for chang-ing the world, must not in principle avoid the (co-)responsibility for its own fate in the world.

The process in the Marxist tradition that is analogous to the "dialectic of enlightenment" could also be analyzed. Furthermore, certain similarities exist between the devastating role of the "dictatorship of the proletariat" in relation to com-munist-humanist ideals and the fateful influence of private ownership in the negative dialectic of liberal humanism with its ideals of freedom, equality, and fraternity.

It is not true, as some would have it, that Marx had in mind only the sociological and not (like the bolsheviks) the political-legal notion of dictatorship as well. There is no doubt that he also pointed out the necessity of *repressive* action by the proletarian state against the overthrown ruling class. And yet, there is an enormous difference between the idea of the dictatorship of the proletariat in classical Marxism and its realization in the bolshevik form. Marx desired and envisaged a transitional dictatorship by the vast majority of the people over the deposed bourgeois minority. But, the ultimate result of bolshevism has been a long-lasting dictatorship by the minority over the proletariat, and generally over the large majority of the people. Exactly how, then, is Marx (co-)responsible for the bolshevization of Marxism?

In my critique of Marx I shall try to proceed as immanently as possible. His ambition was to found a new critical theory. Thus, I begin with a question whose character is metatheoretical: How did Marx actually build his theory and, as its creator, what obligations did he assume, given his ambitions and the criteria he applied to others? Setting out from the criteria of his theory (which should be radically and consistently self-reflexive and self-critical), we can criticize Marx's theoretical procedure for having played into the hands of dictatorial communists (bolsheviks). Marx was a philosopher and thinker of emancipatory practice, a staunch critic of utopian socialism and communism, a great critic of ideology, and a major critic of alienation.

THE UTOPIA OF THE "DICTATORSHIP OF THE PROLETARIAT"

The realizability and the realization of an idea such as the dictatorship of the proletariat should have been the focal point of attention for Marx as a thinker of emancipatory practice and a critic of utopianism. Regrettably, he never raised this problem.

We must include in the core of Marx's theory not only the distant objectives of the communist movement, but also the ways and means by which they were to be achieved, for he wanted to be different from the utopians. We should critically analyze whatever Marx said about these ways and means, but we should also point out the related problems that he did not raise, although he *could and should have done so given the existing level of knowlege and experience in his day.*

If it cannot be realized in a way that preserves its conceived abstract meaning, then perhaps the "dictatorship of the proletariat" can be realized at the expense of that meaning, and itself be (ab)used as an ideological justification for a different practice. Here, then, we are talking not about the *logical consequences* of Marx's idea of the dictatorship of the proletariat, but about the *spectrum of realistic possibilities for its practical application.* This takes us back to my distinction between the ruling and the dominant class.

As we found out in the previous chapter, Marx erred seriously when he embraced entirely different types of class domination within one and the same notion of the "ruling" class. There is an essential difference between a class whose members *structurally* monopolize the state and one whose members admittedly dominate the state but have *no structural* monopoly over it.

Unfortunately, his mistake of looking at every "ruling" class in terms of the structural monopoly model led Marx to the erroneous conclusion that *each and every class rule, even when bourgeois, is more or less class dictatorship.* From here, *per analogiam,* it was just one step to the idea of *proletarian rule and dictatorship.* But, even if the premise of the rule and dictatorship of the bourgeoisie were tenable, this still does not necessarily imply anything about the chances for the proletariat to establish successfully its rule and dictatorship. Since it does not constitute a class of owners of the means of production, the proletariat has an incomparably weaker basis than the bourgeoisie for establishing class domination, let alone class rule and dictatorship.

As an economic (and not a political-economic) class, workers did not (nor could they) become the ruling, let alone the dictatorial class. In order for an economic class (even one possessing only its own labor power) to rule—dictatorially or otherwise—at least one part of it would have to abandon production and become ex-workers. Marx's citing of the Paris

Commune as a model is of no help here:[2] had it lasted, the people would have had to return to everyday affairs. A wise conception of the relationship between classes and state authority ought not set out from a state of emergency.

No matter how we look at it, Marx's idea of the dictatorship of the proletariat was practicable only by having one group rule in the name of the proletariat as a whole. In the best of cases, such a group would rule in the proletariat's interest and under its control. In the worst case, the group would rule without any kind of proletarian supervision and against its vital interests. In conceiving a new state it is no small oversight to set out from the most optimistic assumptions, where no real thought is given to measures and guarantees against the abuse of power. Humankind had plenty of bad experience with its various rulers. These lessons were wisely remembered by liberal thinkers, for instance, and by someone like Jean-Jacques Rousseau, not to mention Mikhail Bakunin's direct warnings, which Marx arrogantly derided. At any rate, by analogy with the differentiation of the Third Estate and the bourgeoisie's separation from it, it was not difficult to imagine the dangers threatening the proletarian class from those excercising dictatorship in its name.

Much has been written about Marx's aggressive attitude toward rivals and critics, and about the impact his attitude had on the development of authoritarian tendencies within Marxism. Marx's personal characteristics are of interest to me primarily as far as they came to expression in the construction

---

2. Ralph Milliband writes: "This points to the fact that the dictatorship of the proletariat in Marx's view of it was meant literally: in other words, that he meant by it not only a form of regime, in which the proletariat would exercise the sort of hegemony hitherto exercised by the bourgeoisie, with the actual task of government being left to others, but also a form of government, with the working class actually governing, and fulfilling many of the tasks hitherto performed by the state." (op. cit., p. 130)

of his theory. We cannot simply shut our eyes to the negative consequences of how a thinker's ideas have been carried out. Moreover, he is *objectively (co-)responsible for the consequences he could have foreseen at the given level of experience and knowledge, but in spite of that did not take into consideration.* This especially applies to a thinker like Marx, who exhibited a great practical intent.

Marx's assumption of the proletariat as a ruling class, and its dictatorship, "answered" all crucial questions before they had even been raised. Had Marx set out from the paradigm of the dominant class, he would have been more prone to contemplate what the eventual dominance of the working class could be based upon. Since this class is not comprised of owners of the means of production, its power over the new state apparatus would have to be based on special institutions and organizations. Unfortunately, within the scope of his class paradigm, Marx was not interested in contemplating their nature or composition.

Nor, as a rule, did Marx's followers look for institutional-organizational means against general statization and the formation of a new ruling class made up of "representatives" of the working class. With Stalinization, the new party and state apparatus did, in the end, become the ruling class. But Marx did not foresee such a possibility and tendency, because he lacked a notion of the political-economic class.

What especially attracted many revolutionaries (not only Bolsheviks) to Marx was his realistic approach to the dangers posed to the revolution by the outgoing ruling class, even when ousted from power. Everyone had heard of the counterrevolution's savage reckoning with the Paris communards. Because of his deterministic approach to the future—after capitalism inevitably comes socialism—Marx did not realize the extent to which the proletariat could be endangered by its "own" state (statism) as well.

It is interesting to note that the *impetus for ideologization* in a theory usually comes with the very first wrong step, the choice of language. My point is not, as some critics claim, that Marx speaks about proletarian dictatorship instead of proletarian democracy. It is to be presumed that those who come to power through revolutionary violence are likely to defend it through dictatorship. Had Marx called his idea *revolutionary dictatorship* he probably would not have put such a lid on the problems stemming from the relationship between revolutionaries and the proletariat, and from the asymmetry between emancipatory objectives and authoritarian means. These problems include, for instance: how to make repressive restrictions affect only the former ruling class, without harming workers; how to prevent the revolutionary elite from becoming autonomous and establishing a dictatorship over the proletariat; how to proletarianize, and then liberalize and democratize the revolutionary dictatorship once it has actually come into being.

Marx's radically anti-utopian self-understanding has long been brought into question. In his picture of a communist future, ample and persuasive analysis has been devoted to those components that bear the signs of utopia, and even of absolute utopia, where all basic contradictions will supposedly be transcended, including the contradiction between man's essence and his existence.[3] However, I take the usual criticism of Marx's utopianism a few steps further and claim that the idea of the dictatorship of the proletariat is itself utopian. Not only is this the case with the ultimate goal, but it is also true of the road leading up to the goal. Out of a number of socialist and communist theoreticians, it is in Marx that the Bolsheviks found support because (among other things) his alleged scientific realism seemed to make him different from the others. Actually, in Marx we have elements of both communist utopia and the utopia of the "transition period." This is not to say, of course, that Marx's entire picture of the new society is utopian.

---

3. I discussed this in chapter 2 of my book: *Between Ideals and Reality* (Oxford University Press, 1973).

Distancing himself from those utopians who countered reality with ideals, and who in the process often appealed to ruling groups and individuals, Marx looked to social life itself for the class vehicle of change toward socialism-communism. But the crucial question that should have been asked in this connection is: does the proletariat have sufficient *power and means* to introduce its dictatorship during the "transition period"? The plausibility of the entire project of the "dictatorship of the proletariat" hinges on the answer to this question, a question of the power and means of basic social groups in capitalist society that is central to the *scientific* approach to socialism and communism. But instead of perceiving and examining this problem, Marx wrongly believed in a kind of *ontological predetermination* of the proletariat:

> It is not a question of what this or that proletarian, or even the whole proletariat, at the moment *regards* as its aim. It is a question of *what the proletariat is,* and *what in accordance with this being, it will historically be compelled to do* [emphasis mine]. Its aim and historical action is visibly and irrevocably *foreshadowed* [emphasis mine] in its own life situation as well as in the whole organization of bourgeois society today. There is no need to explain here that a large part of the English and French proletariat is already *conscious* of its historic task and is constantly working to develop that consciousness into complete clarity.[4]

What strikes the eye is that Marx worked with the concepts of the *being* and the *consciousness* of the proletariat, entirely skirting the question of its class *power,* as though the only problem lay in the development of class consciousness, while the class power of the proletariat is no problem at all. In my view, the question is not "What is the proletariat and what,

---

4. Karl Marx and Frederick Engels. *The Holy Family,* in *Collected Works* (New York: International Publishers, 1975), vol. 4, p. 37.)

in accordance with this being, will it historically be compelled to do?" The real question is: What is the proletariat and what, *in accordance with its being and power, will it historically be capable of doing?*

It is perfectly clear why the development of (self)consciousness was in the forefront of G. W. F. Hegel's system. There would be no point here to any question about the real power of the world spirit, because it ultimately constitutes the cause of all happening. Even someone like Georg Lukács was unable to step beyond the Hegelian framework. That is why he studied the relationship between history and class consciousness, and not between *history, class power, and class consciousness.* In his otherwise self-critical foreword to the new edition of *History and Class Consciousness* (1967), he wrote:

> As to the way in which the problem was actually dealt with, it is not hard to see today that it was treated in purely Hegelian terms. In particular its ultimate philosophical foundation is the identical subject-object that realizes itself in the historical process. Of course, in Hegel it arises in a purely logical and philosophical form when the highest stage of absolute spirit is attained in philosophy by abolishing alienation and by the return of self-consciousness to itself, thus realizing the identical subject-object. In *History and Class Consciousness,* however, this process is socio-historical and it culminates when the proletariat reaches this stage in its class consciousness, thus becoming the identical subject-object of history. But is the identical subject-object here anything more in truth than a purely metaphysical construct? Can a genuinely identical subject-object be created by self-knowledge, however adequate, and however truly based on an adequate knowledge of society, i.e., however perfect that self-knowledge is? . . . Thus the proletariat seen as the identical subject-object of the real history of mankind is no materialist consummation that overcomes the construction of idealism. It is rather an attempt to out-Hegel Hegel. . . .[5]

5. (Cambridge, Mass.: MIT Press, 1968), p. xxii.

We see that even Lukács's self-criticism does not raise the problem of the relationship between consciousness and power, as though the proletariat, even if it were to become entirely conscious of its class interest, would certainly have sufficient power to see it through. And anyway, what does it mean for the proletariat to grow from a class *in* itself into a class *for* itself without soberly taking stock of its own power potential?

One of the most repeatedly discussed questions in the history of Marxism is whether the proletariat can spontaneously develop full class consciousness. V. I. Lenin. as is known, took up Karl Kautsky's idea that revolutionary consciousness should be instilled in the proletariat from "outside." But. if political consciousness has to be introduced into an economic class from outside, then why should not the same apply to its political power? No one, however, has ever mentioned introducing class power into the proletariat from outside!

THE OBLIGATIONS OF A CRITIC
OF IDEOLOGY AND ALIENATION

It is a great pity that Marx did not return to his earlier criticism of "crude" and "despotic" communism. In my opinion, his early works outline what is in many ways a (pre-)Nietzschean genealogy of envious communism that strives for universal leveling. This was a superb model of socio-psychological analysis, but also an analytic framework that had a number of limitations. Why did Marx only mention democratic communism (in contrast to despotic communism), without saying anything about it? Furthermore, are not some people envious not only of wealth and the power it brings, but also of political power as such? How is it that Marx did not foresee a different, more lasting, and dangerous type of despotic communism, one leading not to primitive egalitarianism but, on the contrary, to major social differences, both in political power and in material advantages? Why did it never occur to Marx that leveling despotism (if such a term can be used) would be only a short-lived "transition period" to a despotic rule that establishes privileges for itself and for the strata it needs?

Admittedly, Marx's only experience was with communism in opposition, not with communism in power. The Paris Commune's leveling egalitarianism did not last long enough to give more visible signs of opposite tendencies. It was only later that a form of communism—bolshevism—shifted quickly from the phase of ascetic levelers to the period of Stalinist super-despotism in which the ruling statist class holds enormous privileges.

Experience also teaches us that envy can prompt the "negative abolition" of bourgeois democracy just as easily as it prompted the "negative abolition" of private property. Those who, in the name of the proletariat, try to come to power at any cost also seek to discredit bourgeois democracy at any cost. Such communism never reaches the level of that (or any other) democracy, no matter how much it boasts of having already transcended it.

Why was it that Hegel's analysis of the consciousness of the servant in relation to the master did not inspire Marx to generalize the tendency of not only ruling groups and classes but also the oppressed to develop ideological consciousness? Of course, with Hegel we are talking about stoic, skeptical, and unhappy consciousness. The first simply ignores its subjugated position; the second denies the possibility of any other kind of relationship among people; and the third internalizes its own impotence by turning it into a virtue.

I am referring to a line of thought where Marx's critique of despotic communism could have been more penetrating. It points to yet another form of ideology from below: while invoking the interests of the subjected classes and humanistic ideals, a group can conceal (even from itself) its envy of power-holders and its desire for power at any cost. One wonders whether Marx would have upheld the idea of the dictatorship of the proletariat had he examined it critically in the light of his analysis of despotic communism (and of the so-called Asiatic mode of production and despotism).

Marx was interested primarily in the ideology of the ruling classes, especially the bourgeoisie. He did not integrate his own critique of various forms of communism and socialism into his view of ideology. This only increased the possibility of his theory being (ab)used as an ideology: first as a revolutionary ideology, and later as a ruling ideology. However, as a critic of ideology, Marx was obliged to try, through his mode of theory construction, to reduce the danger of his own thought being ideologized. This is also implied by the principle of critical self-reflection and by Marx's ambition of creating a new, critical social science.

Since, like every future, the future of the communist movement and its theory was entirely open, Marx should have problematized that future. It was not hard to presume that among the tendencies in the movement there would be those who would try, at all cost, to (ab)use Marx's idea for ideological purposes. What would happen if, inspired by The German Ideology and Marx's other critiques of the ideologies of the time, a critique were made of Marx himself? The task of the critic would be to shed as much light as possible on the actual and potential ideological aspects of Marx's work.

In calling for measures to be taken in the construction of critical theory that would strengthen its defense against ideologization, I have no "prescriptions for the future" in mind. Indeed, Marx himself firmly and rightly rejected any such prescriptions. What I do have in mind are broad strokes in elaborating (primarily) the idea of communism as a movement and formulations that are negative and orientational in character. Marx repeatedly stressed that he did not want to construct the picture of the future but to contribute to the self-consciousness of the workers' movement. Would not the said

measures be a contribution to developing the class consciousness of the workers, to transforming them from a class *in* itself
to a class *for* itself?

Many Marxists have made a habit of offering uncritical
and undifferentiated praise for Marx's way of theoretizing on
the future of society. Here is a recent example from an otherwise
good article: "True, in regard to concrete modalities of the
transition to socialism Marx never tied his hands—this is, after
all, an advantage of his concrete-historical way of thinking,
that only anti-historical dogmatists can reproach him for. . . ."[6]
Here, no distinction is made between two things: defining
"*concrete modalities*" of the transition to socialism and showing
the problem in the *basic idea* of that transition.

It would hardly be in the spirit of an emancipatory theory
such as Marx's, which proceeds from the self-activity and self-
organization of the proletariat, to prescribe for the working
class any "concrete modalities." On the other hand, it would
be absurd to suggest that the "concrete historical way of thinking" did not obligate Marx to problematize the *basic principle
of social organization* in the transition period (i.e., the dictatorship of the proletariat). My objection, therefore, has nothing
to do with any construction of the future that would develop
concrete modalities for realizing the idea of the dictatorship
of the proletariat, but rather with the uncritical presumption
of its realizability. History has shown this to be a fatal shortcoming of Marx's theoretical practice.

One can almost coin the following rule: If oppressive consequences can in any way be extracted from an idea, there
will always be a sufficient number of people who will un-
scrupulously try to realize the idea in precisely this way. To
be sure, no thinker can completely prevent the *alienation of
his own ideas*, but he has the obligation to oppose explicitly

---

6. H. Münkler. *Marx heute,* supplement to the weekly "Das
Parlament," Bonn, 12. III. (1983).

and fully such a tendency, especially if, like Marx, he is himself a vigorous critic of alienation.

THE ELEVENTH THESIS ON FEUERBACH AND THE FIRST THESIS ON MARX

There should be no need to remind Marx that in practice (including that of a theoretical nature) not only are existing possibilities realized, but new ones are created. The founding fathers of spiritual orientations, even the greatest and most humanistic, have no right to reserve for themselves an exclusive claim to creative influence, and to shift completely onto their followers the responsibility for the ideological outcome of their ideas and deeds, especially if we are talking about those with an explicit and pronounced emancipatory interest at heart.

Why is it that so many Marxists have taken to making an exception of Marx by reducing his emancipatory interest to his emancipatory intention whenever it suits them? Is emancipatory responsibility restricted to the *intended and predicted* meaning and consequences of somebody's ideas and acts? Marx would never have denied that men, including thinkers, are determined first and foremost by what they actually do, and not by what they wish. Precisely because of the justified effort to overcome subjectivism, the categories of "objective meaning" and "objective responsibility" have been frequently used in the history of Marxism; with full justification, whenever they referred to the whole *predictable* meaning and to all *predictable* consequences of ideas and conduct, but with impermissible ideologization of the notion of responsibility whenever *unpredictable* meaning and consequences were used in order to manipulate people.

Willy nilly, with theories and the way in which they are constructed, thinkers add something to the potential "ideological pool." Social movements, parties, and organizations use these ideas, through selection and transformation, to construct ideologies. This applies also to bolshevism and its relation to Marx.

It is interesting to note the impetus Marx lends, according to G. Cohen, to the study of how ideologies emerge:

> But before an ideology is received or broadcast it has to be formed. And on that point there are traces in Marx of a Darwinian mechanism, a notion that thought-systems are produced in comparative independence from social constraints, but persist and gain social life following a flirtation process which selects those well adapted for ideological service. There is a kind of "ideological pool" which yields elements in different configurations as social requirements change.
>
> Yet it is unlikely that ideas fashioned in disconnection from their possible social use will endorse and reject *exactly* what suits classes receptive to them. Here a Lamarckian element may make the picture more plausible. . . . Because of the delicacy of intellectual constructions, sets of ideas enjoy a partly similar plasticity: one change of emphasis, one slurred inference, etc., can alter the import of the whole.[7]

Of primary interest to me here has been the "Darwinian mechanism" in the process of the bolshevization of classical Marxism. I have been preoccupied with the question of which basic idea of Marx especially lent itself to selection by the bolsheviks, and only mentioned in passing the "Lamarckian" moment, i.e., the bolshevik transformation of that idea. I am not trying to suggest, of course, that Marx, with his theoretical procédé, could have entirely prevented unwanted practical interpretations; rather, I only point out that he did not do nearly enough to narrow down the possibility of at least the most dangerous practical interpretations.

Practical application is not a marginal dimension of Marx's

---

7. *Karl Marx's Theory of History: A Defense* (Princeton, N.J.: Princeton University Press, 1978), p. 290.

theory, and still less is it an external approach to his theory. Marx is renowned for his Eleventh thesis on Feuerbach: "Philosophers have only *interpreted* the world in various ways, the point, however, is to *change* it." One of a number of radically revisionist and critical theses on Marx, should, in my opinion, read as follows: *In order to reduce the danger of the world being changed in an undesired direction, in the name of philosophy, and the philosophy itself being abused as an ideological justification for such change, the way of philosophizing about the world must be changed by focusing on the question of the realizability of that philosophy.*

Resignation, which says that absolutely nothing can be done to prevent the evil fate of theory because there will always be someone to abuse it in one way or another, would be entirely out of keeping with Marx's activist philosophy. After all, the point here has not been to protect Marx's theory from all possible abuse, but from a particular abuse, i.e., dictatorship over the proletariat in the name of the proletariat. Marx wanted to influence and change the world, and to a considerable degree he succeeded. Obviously, he was entirely wrong in predicting the true nature of this influence.

A praxis-oriented thought does not leave the slightest room for giving up on its own fate. In any event, it is far "easier" to influence the fate of one's own theory than to change the world. Also in this regard, Marx would have to reject the Hegelian understanding of philosophy expressed in the metaphor of Minerva's owl, which takes wing only at dusk.

It would be interesting to compare the fatalism of two opposing sides: on the one side there are those who in the name of classical Marxism resolutely reject its bolshevization, while at the same time considering that in the construction of the theory *absolutely nothing could really have been undertaken to prevent this from happening;* on the other side are those who attack classical Marxism, claiming that it *inevitably* leads to bolshevism or its ilk.

Critical rationalism has contributed much to the study of "strategy of immunization" (H. Albert) of theory against criticism. Such a strategy is a pretty good indication of a hidden ideological tendency. What interests me here is the reverse problem, which can eventually open up new areas of fruitful metatheoretical exploration. This is the strategy, so to speak, of immunizing theory against ideologization. I say "so to speak" because there is no chance of creating a theory completely immune to ideologization. But, who says an all-or-nothing stand needs to be taken? There is no reason to underestimate the task of *reducing* the potential of theory for ideological abuse.

I am convinced that just as the full *intellectual potential* of classical Marxism has still not been tapped, so, too, its *ideological vulnerability* has yet to be sufficiently studied and elucidated. Hegel was right in saying that theories are not handed over to followers like money ready to be used; rather those followers must fight for them by dint of their own interpretations. Is it not, however, equally true that the creators of theories are obliged, for precisely this reason, to render those theories as resistant as possible to distortion?

4

Liberal-Democratic and State-Monopolistic Ideology

WHAT IS IDEOLOGY?

Marx's notion of ideology has been systematically reconstructed in the literature many times, thus I would like to make only two points about it. Sometimes he understands ideology as the *idealistic* illusion of self-movement in the world of ideas. This refers to the notion entertained by thinkers, scientists, and politicians that the only genuine impetus for creating and explaining new ideas lies in (existing) ideas. Marx also uses the concept of ideology in another, much more specific way, and, even today, the theory and study of ideology cannot be developed without taking it into account: ideology is a set of ideas that *conceals particular, especially class interests*, presenting them as meta-historical, that is, as God-given, as natural, as universal, and as rational.

In my opinion, ideology should be defined as a *set of ideas that social groups use, at the expense of truth, to justify or discredit a social order or the forces opposing it.* Thus, along with both justifying or descrediting the existing social power, I think that any order put in its place can also have an ideological character. Since Marxism emerged as critique of a given (capitalist) order, Marxist definitions of ideology often overlook this function of *discrediting the existing order* and *justifying the new one*. Conservatives, on the other hand, try to write off as ideological all ideas whose function is to question the status quo and recommend a new social order. My definition avoids this trap as well, because efforts to discredit forces opposed to the existing power structure can also have ideological features. Finally, my definition leaves room for not only the ideology of those in power and those aspiring to power, but also the ideology of those who rationalize their subordinate position.

Anthony Giddens recently came out with some interesting theses on ideology, which he, too, defines in terms of power; but, unlike me, Giddens denies the need to understand ideology in contrast to truth. First he says: "I want to reject the idea that ideology can be defined in reference to truth claims (Thesis I). Then he claims (Thesis II) that the sanctioning of systems of domination suffices as a definition: "Drawing upon this second Marxian strand [which ties ideology to the problem of domination], I therefore propose to interpret the concept of ideology in the following way. I want to define ideology as *the mode in which forms of signification are incorporated within systems of domination so as to sanction their continuance*. I take it to be the typical case of such a notion of ideology that sectional interests are represented as universal interests. This is the basic mode in which forms of signification are incorporated within systems of domination in class societies."[1]

I want to start by making a less important objection to the second thesis, and then move on to cast a critical eye on the first. There is absolutely no reason that, like Giddens, we

---

1. Anthony Giddens, "Four Theses on Ideology," *Canadian Journal of Political and Social Theory* 1-2 (1982).

should limit ideology to *sanctioning* systems of domination, since *discrediting* such systems can have ideological characteristics as well.

In Giddens's view, the concept of 'ideology' is critical rather than neutral. But how can a critique of ideology avoid arbitrariness and relativism if the concept of ideology is not contrasted with truth? For example, let's imagine that a society is attacked from outside and can successfully defend itself only if the social structure is organized along strict hierarchical lines. Can we describe the justification of this order as ideological in the same (critical) sense that we would if it had been allegedly based, for instance, on "God-given" or "natural" human inequality? After all, in the said definition, Giddens himself says that in "basic" ideological cases "sectional interests are represented as universal interests." But, doesn't this mean that ideologies falsely present systems of domination? And how is this to be reconciled with his first thesis?

When using the notion of untruth ("at the expense of truth") to define ideology, I, to be sure, do not mean any naive, objectivistic, or absolutist notion of truth. I agree with Richard Bernstein, who said: "I do not think that there are any fixed criteria by which we can, once and for all, distinguish 'false consciousness' from 'true consciousness'. The achievement of 'true consciousness' is a regulative ideal of the critique of ideology. . . . This does not mean that we must remain intellectually agnostic, that we are never in a position to evaluate and judge the ways in which an ideology is systematically distortive and reflects reified powers of domination. We can show the falsity of an ideology without claiming that we have achieved a final, absolute, 'true' understanding of social and political reality."[2]

There are countless ways in which ideologies try, *at the*

---

2. Richard Bernstein, *The Restructuring of Social and Political Theory* (New York: Harcourt Bruce Jovanovich, 1976), p. 108.

expense of truth, to justify or discredit social orders or the forces opposing them. Only one of them consists of presenting particular interests as universal interests. And there are even completely opposite cases: Nazi ideology, for instance, often attacked conceptions and ideologies referring to universal human interests, and it did so explicitly in the name of racial and national interests. Since Marx's criticism focused mostly on bourgeois-democratic ideology, which belongs to the family of ideologies having universal-humanistic claims, many Marxists unjustifiably define all ideologies as sets of ideas by which particular interests are disguised as universal interests.

Needless to say, my definition of ideology, like any other, can determine only the *primary* function: justifying or discrediting of social orders. Examples of secondary and derivative functions of ideology would include, for instance, stabilization-destabilization or integration-disintegration. It may be superfluous to say that justifying (and/or discrediting), need not be at all direct and open, but can be carried out by diverting attention from or covering up what cannot be justified or discredited, by overemphasis on positive or negative aspects, and by other methods.

Those who *define* ideology as "distorted" or "false" consciousness forget that there are many forms of such consciousness which have no ideological character at all. They are quick to assess some particular kind of consciousness before anything is said about its properties and functions. *Distorted, false,* and (as I have suggested elsewhere) *lying* consciousness are, in my opinion, dominant dimensions of ideologies, or at least significant phases in the transformation of ideologies.

Ideology as "distorted" consciousness is a mixture of truths and untruths, and as "false" consciousness it is a set of untruths. The difference between "false" and "lying" consciousness is that in the first case the proponents of ideology are not aware of its untruthfulness, whereas in the second case they are. Ideologies usually emerge in history as "distorted" consciousness.

With time, as they exhaust themselves, they increasingly—*consciously*—try to defend themselves at the expense of truth.

It may seem to the reader that my suggestion of "lying" consciousness merely constitutes a return to the Enlightenment's obsolete approach to ideology. As is known, the great thinkers of the Enlightenment interpreted and rejected dominant medieval ideas as *conscious* (i.e., deliberate) deception in the service of both the state and the church. This was too simplified a view for Marx and Engels: they spoke about "distorted" or "false," but not "lying" consciousness.

The history of philosophy and social thought has taught us one important lesson: conceptions that appear to have been buried forever often re-emerge, albeit with important modifications and much more limited application. In my opinion, this is the case with the Enlightenment's concept of ideology. *Sometimes, ideologies predominantly use conscious lies.* I am sure, for instance, that Joseph Göbbels and his Nazi propaganda apparatus knew they were lying in their ideological diatribes. For this reason we should adopt the Enlightenment's insight about "lying" consciousness, but, at the same time, should limit it to a dimension or phase of ideology.

In the following passage from Marx, he comes very close to such a view:

> The more the normal form of intercourse of society, and with it the conditions of the ruling class, develop their contradiction to the advanced productive forces, and the greater the consequent discord within the ruling class itself as well as between it and the class ruled by it, the more fictitious, of course, becomes the consciousness which originally corresponded to this form of intercourse (i.e., it ceases to be the consciousness corresponding to this form of intercourse), and the more do the old traditional ideas of these relations of intercourse, in which actual private interests, etc., etc., are expressed as universal interests, descend

to the level of mere idealizing phrases, conscious illusions, deliberate hypocrisy.[3]

One reason this idea was not expressed in Marx's definitions of ideology would seem to be that he concentrated on analyzing and criticizing the bourgeois ideology of his day and age, which can freely be described as "distorted," although certainly not as "lying" consciousness. In any case, that was a time when progress in understanding the ideological phenomenon truly required distancing oneself as far as possible from the Enlightenment's oversimplifications.

WHAT SHOULD IDEAL-LOGY BE?

There is yet another dimension or phase of ideology that should be introduced—ideal-logy. I first used this term in my book *Geschicte und Parteibewusstein*.[4] I would now define ideal-logy as *a set of ideals that social groups use, at the expense of truth, to justify or discredit a social order or the forces opposing it.* The tying up of ideal-logy with untruth ("at the expense of truth") provides for the distinction between the nonideological formulation and justification of ideals, on the one hand, and ideal-logy on the other.

But ideals cannot be distorted or false in an epistemological sense. And yet, this dimension or phase of ideology is also characterized by subordinating and sacrificing truth to the requirements of justification or discrediting. The principal modes of ideal-logical (self)delusion are yet to be systemically studied.

---

3. *The German Ideology, Collected Works Volume 5* (New York: International Publishers, 1975), p. 293.

4. (München, 1978), p. 142. An English translation was published in 1981 by Prometheus Books, (Buffalo, N.Y.) under the title *In Search of Democracy in Socialism*.

In ideal-logy, untruth (partial or complete) is often expressed when comprehending the relationship between means and ends, especially the ultimate ends (ideals). Ideal-logies most often set social goals, and only then seek out suitable means to realize them. With their ideal-logies, social groups often delude themselves and others as to the realizability of their goals and ideals, the nature of the means they use, and in terms of believing that they can realize their declared goals and ideals through *these* particular means. How else can one fully understand and analyze futuristic ideologies, especially radical ones? Even important intellectuals can be the victims of an ideal-logical approach to reality. The history of communist and other parties is full of such examples.

In ideal-logies, the nature of the means looks as though it is strongly determined by the ends, even the most distant, whereas in reality the relationship between means and ends is most often the other way round. Unlike social ends, social means *also* exist outside the consciousness of the actor (individual and collective) and so, in the long run (and ultimately), they are, as a rule, ontologically superior to ends, especially the supreme ends. It is not wonder then that political activity so often witnesses an inversion of means and ends: the former become ends unto themselves and the latter serve as their ideal-logical justification.

Revolutionary ideologies counter the existing order with an ideal, i.e., a radically different social system. It is in this predominant form—of ideal-logy—that bourgeois-democratic ideology appeared on the historical stage. This is how communist ideology also embarked on its road. Ideologies most often begin their life as ideal-logies, and then more and more are transformed into *real-logies*, a term that I have proposed elsewhere to designate a special dimension or phase of ideology. I have in mind a spectrum with ideal-logy and real-logy at either end. In order to justify a social order (or the forces opposing it), *real-logy no longer refers to ideals, but rather to their "realization" and the principle of "realism."*

Bourgeois-democratic ideology has traversed the road from conceiving democracy as the "rule of the people" to actually reducing it to pluralism and competition among political elites for power. This real-logy has brought about a certain de-ideologization of bourgeois democracy. However, at the same time it has been proclaimed as not only the greatest achievement in the development of democracy to date, but also as its ultimate possible achievement—and this, again, is ideological (self)delusion. Surprisingly, no one has coined a term that captures this negative legitimization in contrast to "real socialism": *real democracy*, allegedly, can only be the circulation of ruling political elites by means of free elections, and it can succeed only in capitalism.

Bolshevism has also gone through an ideal-logical phase primarily invoking the ideal of a classless and stateless society. The transformation of the Bolshevik ideal-logy into the ideology of "real socialism" (a kind of real-logy) will be the subject of chapter 6.

ERRONEOUS GENERALIZATIONS

The history of the twentieth century has been strongly marked by the clash between two major ideologies: communism and bourgeois-democracy. Their *structural tendencies* are very different, even contrary to one another. That is why any attempt to take one or the other type of ideology as a model for generalizing and defining the concept of ideology leads one astray. Let me illustrate this by two unsuccessful attempts, one from the anti-Marxist and the other from the Marxist circle.

Karl Dietrich Bracher, the West German historian and theoretician, takes the following approach to ideology:

> The character and workings of modern ideologies can best be seen in the sharpest, most exclusive form, which emerges when

totalitarian political objectives and styles of thought step onto the scene. The essence and function of ideologization in the state and society are especially reflected in totalitarian views of the world, whether we ascribe them to older strata of monocratic thought or extract them from Rousseauism and the radical egalitarianism of the French Revolution, or want to ascribe them only to the Left-radical and Right-radical extremism of socialism and national-socialism in our century.

The nucleus of this process is the tendency to oversimplify complex realities: the striving to reduce them to a *single* truth and at the same time to dichotomize them into good and bad, right and wrong, friend and foe, to view the world in a bipolar way on the basis of a single model of explanation. Marxist class and national-socialist racial theory are especially two cases in point. Building stereotypes of the enemy and a strategy for seeking out the sacrificial lamb as means to simplify and integrate social plurality is as important as are vague promises and visions. . . . Guarantees of absolute truth, not just in heaven but here on earth, give ideology the character of a secularized religion of salvation and atonement.[5]

Like Bracher, many Western analysts and critics of ideology have been so "impressed" by the totalitarian kind of ideologies (national-socialist and Stalinist) that they have proclaimed their properties as universal. Hardly anything of what has been quoted from Bracher can be applied, for instance, to bourgeois-democratic ideology. Is not the designation "ideology" itself ideologically biased when it does not also embrace this major and still very powerful ideological complex?

All the features that Bracher sees as the defining characteristics of ideology are for Agnes Heller indications that this is not ideology in the true sense of the word. In her opinion, *each and every ideology* claims to "represent the cause of the whole of mankind," yet despite this universalistic

---

5. "Zeit der Ideologien," Stuttgart, 1982, p. 168. (My emphasis.)

pretension it "only generalizes class-specific, i.e. particular interests."[6]

Such a restrictive definition was bound to lead to an unacceptable conclusion: "According to this definition we reach the conclusion, which may sound bold, that so-called Soviet ideology is not really an ideology at all."[7] And, why is this alleged "Soviet ideology" not to be included in ideologies, but only in "state doctrines" and "systems of dogmas"? Heller gives three reasons:

(1) Ideology (according to the above definition) pretends to express universal interests (and in fact it expresses class interests) whereas "Soviet ideology" contends to represent class (proletarian) interests. However, we have already established that there are other ideologies that also do not pretend to express universal interests.

(2) "Ideologies in the above sense compete, emerge on the market, always see themselves as part of plurality. Marxism-Leninism, on the other hand, does not join any race; it excludes all other ideologies. And thereby it ceases to be an ideology in the real sense of the word."[8] The very mention of competition in the market shows that it is the liberal ideology that here serves as a model for definition. Thus the ideological field is further narrowed down: not a single state-monopolistic ideology—and they unquestionably account for the majority both in the past and today—is really an ideology.

(3) The Soviet "state doctrine" is not concerned with its own coherence since, like a creed, it is not subject to criticism. This introduces further narrowing down of the concept of ideology, this time from a rationalistic-logical perspective. The result: the vast majority of past and present ideologies could not be encompassed by this concept.

---

6. A. Heller, F. Feher, and G. Markus: "Der Sowjetische Weg. Bedürfnisdiktatur und entfremdeter Alltag," VSA, 1983, p. 217.

7. Ibid.

8. Ibid., p. 218.

Heller warns that she does not believe that "there has never been ideology in Soviet-type societies." According to her, it took some time to transform "Marxist ideology" into "state doctrine." It was not until around 1922 that non-Marxist theories started being banned. For a while members of the "party aristocracy" had an equal right to take part in debates on the common theoretical heritage, and in mutual discussions they continued to use rational arguments. Later, true Marxism became illegal precisely because (through its "ideological character") it had allowed and brought a certain "theoretical pluralism." Thus, it would turn out that only non-state *ideology* exists in "Soviet-type societies"; that Leninism is ideology, but Stalinism is not; that, for instance, somebody like Lukács is an ideologist, but Stalin is not.

I would like to point out, however, that I have only criticized Heller's definition of ideology and not other aspects of her analysis of the "Soviet state doctrine."

TWO OPPOSITE IDEOLOGIES

Bourgeois ideology, both in its economic-liberal and in its political-democratic forms, draws and *limits* attention to what is evident: competition, equal rights, elections, political pluralism, and so on. Since it is largely rooted in everyday experience, this ideology is "capillary" in the way it spreads and has a great magnetic power. However, in fixating our consciousness to immediate reality, bourgeois ideology also conceals the mechanisms of class domination.

No one needs to try hard to hide these mechanisms: they are *structurally* concealed. The separation of "civil society" and the state offers room for legal and political freedom and equality, as well as democracy, and, at the same time, this separation makes the class domination—which is carried over from the economy into other spheres of social life and power—hardly visible. It is perfectly understandable that the bourgeoisie (as the nonruling dominant class) has comparatively less of a need to engage in centralized ideological production and indoctrination than does the statist (ruling) class, though at first glance we might expect it to be the other way round, because the statist class, in addition to its ideological monopoly, can, if necessary, use its monopoly over state repression.

Marxists who have forgotten the separation of "civil society" from the state in capitalism overlook the fact that bourgeois ideologies are emerging, spreading, and operating primarily in the sphere of "civil society." Just as the birthplace of the dominant class in capitalism is in this sphere, and not in the state apparatus, so bourgeois ideologies have their birthplace in that sphere.

This is not to say that Gramsci erred when he stressed the role of intellectuals and ideological apparati (educational institutions, the mass media, and others) in capitalism. But, "ideological apparati" and "ideological state apparati" (Althusser's expression) are not the same things. It is absurd, as Althusser has done, to treat the family, the entire school system, the church, and all mass media as ideological *state* apparati. In capitalism not even political parties are state organizations but rather private entities.

However, even if we were to apply the notion of "ideological state apparatus" properly, we could not ascribe to it such an immense and, still less, central role in creating and spreading bourgeois ideologies simply because they originate in "civil society" rather than in the state apparatus. In overlooking the separation of the "civil society" and the state in capitalism, it appears that Althusser unconsciously projected into this state the place and function of ideology in a typical ruling communist party (here the category of ideological *party-state* apparatus would fit). Anyone who uses as his model the way in which ideology is created and spread in countries where communist parties are in power, and then applies that model to the West, will never really grasp how ideology emerges and is spread in capitalism.

To be sure, liberal and democratic bourgeois ideology has the form not only of the everyday and spontaneous consciousness of the "ordinary" man, but also of "high" ideology. Since in terms of content there is a homology between these two ideological levels, the latter falls on the fertile ground of the former (of course, the everyday and spontaneous level is fragmented, incoherent, and sometimes even chaotic). However, one cannot talk about "introduction of consciousness from outside," but rather about systematization, elaboration, and justification of the existing everyday consciousness. The dominant class of "civil society" in capitalism has as a rule no (single) "vanguard party" that pretends to interpret and represent its "objective interests."

It is an entirely different matter, however, when we talk about an ideology whose basic ideas, according to the Bolshevik project, are created outside of everyday and spontaneous consciousness of the working class, and need to be *introduced into it from the outside* (even despite its *resistance*). It is not surprising, therefore, that bolshevized communists in the West have never fully understood why liberal and democratic ideology have a much greater influence over "ordinary" people, and even workers, than their ideology of the "objective interests" of those same people and workers. The Bolshevik approach is fundamentally wrong because it postulates that liberal and democratic ideology must also be introduced in the working class *from outside*.

Practically all the basic differences between these two types of ideologies can be expressed by the image of market competition on the one side and monopoly on the other. Market competition and selection is said to be the basic mechanism regulating relations between the economic, political, and ideological contestants in capitalism.

State-monopolistic ideology, of course, does not allow for

the market of ideas. Free supply and demand may eventually exist regarding material goods, but not with respect to spiritual products of key importance to achieving ideological control. One part of the Leftist intelligentsia with its sweeping, undifferentiated attack on the "market in culture," has unwittingly helped to suppress the autonomy of intellectual work in statism.

An ideology in which all (even merely potential) competitors are declared enemies suits a centrally planned, distributive, and command economy. This is a question of the manichaean polarization of the world. A kind of permanent "war communism" is marked by an irreconcilable "ideological struggle." This is a language overflowing with wartime metaphors: front, battlefield, battle, and the like. It takes a lot of time, social resistance and crisis before such a "vanguard" realizes that, with no major threat to itself, the slogan "Who isn't with us is against us" can be replaced with the slogan "Who isn't against us is with us."

This manichaean mentality has carried over to emigrants as well. Here is what one of them, Andrei Sinyavski, said in an interview with a West German paper:

> Roughly and schematically speaking, one can now distinguish between an authoritarian-nationalistic and liberal-democratic wing of Russian emigre dissidents. The leader of the first group, which is supported by the majority of emigrants is Alexander Solzhenyitsin. I count myself in the second group.
>
> The gap between those in Russia who think differently is traditionally deep. For us someone who thinks differently than we do is our enemy, "the agent of imperialism"; here they call him "the agent of the KGB." The Soviet authorities have drummed into our heads a vision of omnipresent enemies, enemies, enemies against whom we must fight, fight, fight.[9]

---

9. *Die Zeit* (October 19, 1984).

Characteristic of *liberal*-democratic ideology is that it allows activity by every group and party that respects the state constitution. However, some other ideologies in capitalist countries (under certain conditions even the dominant ideology) have structural tendencies that are very similar to the dictatorial-statist ideology. Suffice it to recall the manichaeism of McCarthyism or the current fundamentalism of the "moral majority" in the United States, which is obsessed with both the "communist threat," and the fear of a liberal attitude to the "communist enemy."

The image of an open or occupied space can also help us to compare the two ideologies. "Market" suggests the image of a space accessible to everyone in principle. State monopoly ideology, on the other hand, has a need to homogenize and fill in the ideological space. As soon as a crack appears, ideological guardians rush to eliminate all "foreign" ideas. They are panic stricken by the "empty space the enemy can use." Without this striving for totality, one cannot understand the big *surplus of ideology in statism*. Because of the growing *ennui* it produces, this kind of "Marxism" becomes increasingly repugnant.

While the liberal-democratic ideology is characterized by its abstraction from differences, by blunting conflicts, and by neutralization, dictatorial-statist ideology aggravates conflicts and demands universal allegiance to the party in the name of "science." Lenin himself took pride in "scientific ideology." This ideological monopoly implicitly justifies itself by claiming that its challengers would ultimately have to be "laymen" who have no place in a "scientific debate." There is also a kind of "scientific" monopoly over definitions: they are seen as empirical statements, even scientific laws, not as stipulations by ideological power holders. There has long been a separate discipline in the Soviet Union, studied and taught at all levels: it is called "scientific communism." I would say it is a kind of social science fiction. From its emergence to the phase of "real socialism," this ideology has faced almost insurmountable obstacles in understanding real capitalism. Since social science and philosophy in the USSR are not free, the "Soviet understanding" of the West depends much more on the interests, experience, and mentality of Soviet leaders than is the case with the Western understanding of the USSR.

It would be very important to study the hermeneutic aspects of the confrontation between liberal-democratic ideology and the state-monopolistic ideology. There is a whole range of self-projections that impede them from properly understanding each other. An analysis of this social-psychological mechanism could, in my opinion, open up major prospects for a comparative study of ideologies.

What primarily interests me here is the fact that a ruling (statist) class projects its power structure onto a capitalist system where there is no ruling class, but only a dominant one. It is no wonder that statist ideologists misapprehend the dominant ideology in capitalism, since these two mutually opposed ideologies function, spread, and gain influence in entirely different ways. Here we are dealing with social totalities of which the one is dominated by *monopoly politics* and the other by *competitive economics*. It is extremely hard for an ideology marked by monism, monopoly of party-state apparatus, monological structure, scientific-nomological pretensions, and repression, to fathom the ideology of pluralism: the circulation of elites in governing the state, the use of dialogue, the separation of science from ideology, the attachment to social contract, and the belief in tolerance.

ON MARCUSE'S CRITIQUE OF "REPRESSIVE TOLERANCE"

Marcuse concentrates[10] on analyzing the historical development of tolerance: from liberating tolerance at the beginning of

---

10. In his essay "On Repressive Tolerance," in R. P. Wolff, B. Moore, Jr., and H. Marcuse, *A Critique of Pure Tolerance*, (Boston: Beacon Press, 1965).

modern times to the present day.[11] He says that today tolerance
is an integral part of the ruling ideology because it obliges
respect for the rules of hierarchically structured power. To put
it ironically: it is slightly harder for those at the bottom to
tolerate those at the top of the hierarchy of power, than the
other way around. Delving through the veil of abstract ideas-
values, Marcuse tries to reach the source of concrete power,
and there discover the "background limitations." His findings:
What seems like unbiased tolerance actually shields the
"established machinery of discrimination" and suits conservative views and movements, discriminating against all Leftist
and generally progressive forces.

Marcuse's syntagma of "repressive tolerance" calls attention to the need to examine the ways in which tolerance
is transformed into the ideology of domination. Such
syntagmata are suitable for heuristic purposes, because they
draw attention to the limitations in realizing values and even
to their functional transformation into their own opposites.
In such a case, one should talk about the *negative dialectics
of tolerance.*

The trouble does not start until such *initial and heuritstic
moves are seen as final theoretical categories.* Marcuse should
have drawn a sharper line between tolerance and repression,
rather than almost erasing it with the notion of repressive
tolerance. As a result, he, unfortunately, played into the hands
of ideologists of his own theory. This example shows the major
role language plays in creating the ideological potential of
theory. The notion of *manipulative tolerance,* in my view, would

---

11. Sometimes one gets the impression that Marcuse compares
the ideal(-logy) of tolerance in the early development of capitalism
with the practical fate of tolerance in more recent times. It is not
true, however, that there was more tolerance earlier in capitalism:
the right to vote was highly limited at the time, political and trade
union organizations of workers were not allowed, etc.

have been much less open to the danger of ideologization than the concept of "repressive tolerance."

There have been Leftists who turn Marcuse's critique of the "ideology of tolerance" into an *ideology of "repressive tolerance."* They try to discredit democracy at any price, even at the price of truth. Proceeding from the correct premise that differences between democracy and dictatorship in capitalism are not absolute, they draw the wrong and very dangerous conclusion that these differences are negligible. This is one of the countless variations on the claim that the differences between tolerance and democracy on the one hand, and repression and dictatorship on the other, are only *formal* in character, whereas in *essence* in both cases it is a matter of the dictatorship of the ruling class.

What, according to Marcuse, is repressive tolerance to be replaced by? Selective intolerance: expanding tolerance for ideas and movements on the Left, and intolerance for those on the Right. He goes so far as to recommend not only "the withdrawal of tolerance of speech and assembly from groups and movements which promote aggressive policies, armament, chauvinism, discrimination on the grounds of race and religion," but also from those "which oppose the extension of public services, social security, medical care, etc.," and, generally speaking, "by their very methods and concepts, serve to enclose the mind within the established universe of discourse and behavior—thereby precluding *a priori* a rational evaluation of the alternatives."[12] The scope of intolerance is thereby extended to the "stage of action as well of discussion and propaganda, of deed as well as word."[13]

Marcuse feels, of course, that the following two questions are of crucial importance here: (1) Does an *objective criterion* exist to differentiate what should from what should not be

---

12. Marcuse et al., *A Critique of Pure Tolerance*, p. 100f.
13. Ibid., p. 109

tolerated? (2) *Who* is competent to apply such a criterion? He says this criterion consists of "reasonable chance for pacification and liberation." A "historical calculus of progress" is possible, which should be based on rationality comparing the consequences of maintaining the existing state with the consequences of supporting the alternative. Unfortunately, Marcuse does not analyze the historical experience showing that movements, including those on the Left, which substantially restrict freedom of speech, assembly, and activity seldom lead to "pacification and liberation." Does not the rational basis for assessing what we can expect from victorious social movements include their attitude—be it guardian or destructive—toward the scope already won for democracy and tolerance?

Marcuse's answer as to who is competent to make the assessment is equally poor: "Everyone 'in the maturity of his faculties' as a human being, everyone who has learned to think rationally and autonomously. The answer to Plato's educational dictatorship is the *democratic educational dictatorship of free men.*"[14] This answer only multiplies the problems. Who should separate those with "mature faculties," who think "rationally and autonomously," and who are thus "free," from other people —and how? How could one prevent what happens so often, i.e., dictatorship in the name of freedom turning into dictatorship against freedom? Finally, what does "democratic educational dictatorship of free men" really mean? What would it roughly look like? How would it be introduced? What strikes the eye is that Marcuse's idea is even more vague than Marx's notion of the dictatorship of the proletariat.

Marcuse says not a word about these key questions in his essay "On Repressive Tolerance" or practically in any of his other writings. Hence, I have to say that he (in spite of his unquestionable contributions to the development of humanistic Marxism), having facilely launched the idea of "the democratic

---

14. Ibid., p. 106. (Italics mine.)

educational dictatorship of free men," laid the groundwork for the possible ideological abuse of his own conception. It is yet another occasion to reflect on the social responsibility thinkers assume for the way they construct their theories.

In this context Marcuse's negative stand on dictatorial communism is not much help.

> The factual barriers which totalitarian democracy erects against the efficacy of qualitative dissent are weak and pleasant enough compared with the practices of a dictatorship which claims to educate the people in the truth. With all its limitations and distortions, democratic tolerance is under all circumstances more humane than an institutional intolerance which sacrifices the rights and liberties of the living generations for the sake of future generations.[15]

Marcuse does not even raise the question of how to avoid the danger of his "democratic educational dictatorship of free men" being deformed into such a dictatorship. He seems to have realized the difficulties, and only three years later in a 1968 *Postscriptum,* said something that is rather different:

> However, the alternative of the established semi-democratic process is *not* a dictatorship or elite, no matter how intellectual and intelligent, but the struggle for a real democracy.[16]

But Marcuse's attitude toward democracy in highly developed capitalist states remains quite ambivalent. He criticizes this democracy as "totalitarian" and as a "tyranny of the majority," but also as the class rule of the bourgeoisie. However, if the bourgeoisie rules, there cannot be genuine, but only seeming tyranny of the *majority.* Here is further proof of Marcuse's ambivalence:

---

15. Ibid., p. 99.
16. Marcuse et al., *A Critique of Pure Tolerance* (Boston: Beacon Press, 1968), p. 122f.

The exercise of political rights (such as voting, letter-writing to the press, to senators, etc., protest-demonstrations with a priori renunciation of counter-violence) in a society of total administration serves to strengthen this administration by testifying to the *existence of democratic liberties which, in reality, have changed their content and lost effectiveness.* In such a case, freedom (of opinion, of assembly, of speech) becomes an instrument for absolving servitude. And yet . . . the *existence and practice of these liberties remain a precondition for the restoration of their original oppositional function,* provided that the effort to transcend their (often self-imposed) limitations is intensified.[17]

But, how can something that has already lost its effectiveness be the precondition for any kind of new effectiveness? One gets the impression that Marcuse, like many other Marxists, often does not know what to do with democracy in capitalism.

True, two deep experiences lie at the root of Marcuse's reaction to the social situation. They are totalitarianism (both in its Nazi and in its Stalinist versions), on the one hand, and American democracy, with its phobia of any kind of socialism, on the other hand. Still, Marcuse's advocacy of a "democratic educational dictatorship of free men" is, in my view, unacceptable not only to liberals but to democratic Leftists as well. To be able to judge the practical-political significance of Marcuse's basic idea in the essay analyzed here, it would be helpful to remind oneself as well that in democratic capitalism the only successful communist parties are those whose members fight to expand tolerance rather than selectively introducing intolerance.

---

17. Ibid., p. 84. (Italics mine.)

5

Capitalism, Statism,
and the Critique of Ideology

MARXIAN CRITIQUE OF BOURGEOIS IDEOLOGY

Marx's critique of capitalist ideology is based on contrasting "form" and "content," "appearance" (*Erscheinung*) and "essence," and finally "appearance" (*Schein*) and "reality." Marx borrowed these conceptual pairs from Hegel and adjusted them to his own purposes. In my opinion, he was most successful in applying them to *economic exchange* as the main source of the *ideology of economic liberalism*. Marx tried to unmask the equality and freedom of *purchaser and seller* as a form and appearance (*Erscheinung* and *Schein*) that conceals the true content, essence, and reality in *production*, i.e., the bourgeoisie's domination over workers. Admittedly, noneconomic coercion no longer exists as it did in feudalism, but it is for Marx no less true that the worker is economically dependent since he is forced to sell his labor to the bourgeoisie.

Marxists, as a rule, apply this model of critique to another form of capitalist ideology as well: democracy. Now it is a question not of the freedom and equality of citizens as purchasers and sellers, but as voters. Marxists want to pierce through the *political-legal* veil over bourgeois domination, which, in their opinion, is inevitably transferred from the *economy* to the equality and freedom of citizens vis-à-vis the state, thus turning them into form and appearance (*Erscheinung* and *Schein*).

Marx established an important theoretical paradigm for a critique of capitalist ideology. The Bolsheviks, however, ideologized this paradigm. Marx's paradigm prompts us to look for hidden class domination, since "form" *abstracts from more concrete content,* and "appearance" (*Erscheinung*) does not overstep the framework of *obvious and unmediated reality.* However, in their critique of bourgeois democracy, "form" for Marxist-Leninists, and even more for Leininist-Stalinists, means *outward shape,* which is more or less irrelevant for determining content, and "appearance" (*Erscheinung*) is only the *surface and unimportant side of an entity.* The worst consequences in the history of Marxism have been produced by the simplified and deformed application (as *pure illusion* instead of as a combination of truth and untruth) of Hegel's category of *Schein* (appearance) to freedom, equality, and democracy in capitalism.

The transformation of Marx's theoretical paradigm into an ideological one, and the underestimation of the possibility of democratizing capitalism, resulted to some extent from simply carrying over criticism of economic liberalism to democracy. However, just as economic liberalism does not automatically lead to democracy (rather the citizen must fight to win it), so the Marxist critique of economic liberalism cannot automatically be extended to democracy in capitalism. At first, it seems as though this can be done without any major modifications, since in both cases it is a question of ideologically equalizing people (freedom and equality) by means of abstracting from the concrete class position: in the first case, we are dealing with the purchaser and seller of labor power, and in the second with the voter.

But, is not the following difference sufficiently indicative: while the one sells and the other buys labor power, everyone is a voter. To be fair, in Marx's time there were property as well as educational and sexual electoral barriers, and it could have seemed as though de-masking "freedom" and "equality" in the sphere of circulation might simply be extended to the electoral process. However, ever since democracy-for-all-citizens has come into being, the category of economic compulsion, otherwise welcome in the criticism of economic liberalism, leads us astray rather than indicates the true limitations of such democracy in capitalism. For while workers *have* to sell their labor power in order to live, they *may* vote or not, vote for the candidates of workers' parties, or exercise the franchise in other ways.

In saying this, of course, I do not deny that even in the most democratic state the *influence* of the bourgeoisie can be dominant. Rather, I am suggesting that its domination is much more subtle (than suggested by the notion of economic compulsion), more indirect than direct, and more negative than positive. In other words, the question is not one of economic *compulsion* over workers as voters, but one of economic and other *limitations* primarily for carrying out a radical program if such workers' parties gain control of the government (which, let us not forget, is only a part of power) through parliamentary means.

Why is it that many Marxists still fail to see that since Marx's time the focus of capitalist ideology has increasingly moved from the sphere of economic circulation to the sphere of politics?[1] They still write critical discourses primarily on

---

1. And in the sphere of economic circulation—to consumption. It is an ideology of "freedom" and "equality" of workers not so much as sellers of labor power but as buyers of consumer goods.

economic liberalism, believing that this is the most efficient way to attack capitalism. But democracy has long since been the main pillar of capitalism's ideological legitimation. The power structure in capitalism was more transparent while democracy was in swaddling clothes. The more democratic capitalism is, the less transparent is the class domination over the state. Therefore, one has to tackle capitalism on this very same terrain, and convincingly show how and to what extent democracy is *limited* by the capitalist mode of production and domination and how these limitations might be removed.

Thus far my analysis and critique have been immanent. In other words, I have not been raising doubts about Marx's conceptual apparatus—form-content, appearance (*Erscheinung*)-essence, and appearance (*Schein*)-reality—but have only examined its applicability to democratic capitalism. It is these concepts, however, that have been increasingly subjected to critical scrutiny even by Marxists or scholars sympathetic to Marx.[2] According to these objections, the metaphysical tradition—Plato, Aristotle . . . Hegel—greatly influenced Marx. Hence, there are more than traces of essentialism and ontologism in Marx. It seemed to him that there is a dimension of reality that is at once hidden and epistemologically privileged, a dimension that has to be penetrated in order to find the clue to the riddle of historical development. That (so to say) Archimedian point Marx believed he found in the mode of material production. Here is how John Kean summarizes the basic objection to Marx:

---

2. Here I would like to mention only G. A. Cohen's book *Karl Marx's Theory of History: A Defense* (Princeton, N.J.: Princeton University Press, 1978); an article by Milan Pruha, "The Crisis of Marxism as a Limitation of an Emancipation Theory" (SerboCroat, *Theoria*, 1-2 [1983], Beograd); and an essay by John Kean, "Democracy and the Theory of Ideology," *Canadian Journal of Political and Social Theory*, 1-2 (1983).

The classical Marxian critique of ideology is nowadays reproached with having reproduced a certain form of naturalism, inasmuch as it consistently avoided questions concerning language, or, more precisely, the crucial importance of signifying practices within social life. . . . According to Laclau, Hirst, Pecheux, Gadet and other critics (most of whom are indebted to Althusser), the classical Marxism denoument of the riddles of ideology rests upon the misleading and untenable distinction between the ideological forms in which "reality" appears or presents itself and a prior ontological domain of "reality" which consists of "material" life activity ungoverned by processes of signification. As a consequence of this distinction, ideology is understood as a form of posthumous misrepresentation of a subterranean reality of material life processes; conversely, these material life processes are interpreted as the presymbolic point of origin which is also the point of truth that contradicts the "false" dissimulations of ideology.[3]

True, in any critique of Marx the following point, underlined by William L. McBride, has also to be taken into account:

"Essence" is not an ideal choice of word, since it means something different, something more metaphysical, in the writings of both Hegel and Aristotle, than it does in its very occasional employment by Marx. Marx is not referring to allegedly eternal essences, but rather to underlying mechanisms, detectable only after lengthy analysis, within an existing system in particular, within the capitalist system that most interests him. Still no other alternative choice of words conveys the idea much better.[4]

In spite of this, however, the contrast of form-content, appearance (*Erscheinung*)-essence, and appearance (*Schein*)-reality

---

3. John Kean, "Democracy and the Theory of Ideology," *Canadian Journal of Political and Social Philosophy* 1 (1983): 8.

4. *The Philosophy of Karl Marx* (New York: St. Martins, 1977), p. 68.

has to be understood as being much more relative, context-dependent and purpose-dependent, and ultimately as being much more arbitrary than it seemed to Marx and many Marxists. Thus, "freedom" and "equality" of sellers and buyers of labor power ought to be characterized as *formal* and *apparent* when they disguise bourgeois domination. On the other hand, the very same freedom and equality belong to the structural features of capitalism and should, accordingly, be termed substantial, essential, and real, because without the freedom and equality of the buyer and seller of labor power capitalism would be inconceivable.

If this is so, however, is it not also true that there belong to the *essence of capitalism* not only bourgeois domination but such freedom and equality as well? At any rate, sellers of labor power have used their freedom and equality to diminish and limit bourgeois domination, to improve considerably their own position, and even to become citizens who possess equal rights, universal suffrage included. Hence the following theoretical ramification: the understanding of "essence" should be entirely dialectical since it is not homogeneous but contradictory.

Is not the contradiction between equal rights of citizens and bourgeois domination one of the *essential characteristics of democracy in capitalism*? Equal rights of citizens do not represent here only form and appearance. After all, workers have successfully benefited from their power as equal citizens, using it to increase their power as producers. Is it not misleading to characterize such powerful levers as these as just belonging to the sphere of form and appearance, without at the same time explicitly and fully pointing out the relativity of such formulations?

The greatest weakness of Marx's category apparatus was the underestimation of the reformist and democratic potential of capitalism. True, Marx did occasionally, even radically, overstep his conceptual framework in his *ad hoc* judgments (although only implicitly): for instance, when he allowed for the democratic-parliamentarian transition to socialism in some capitalist countries. However, it is highly questionable how something purely formal and apparent can, at the same time, be seen as a lever even for a wholesale systemic transformation and not simply for the transformation of some aspects of the system. Unfortunately, Marx never reexamined his own view of the bourgeoisie as a ruling class, and of freedom, equality, and democracy under capitalism as forms and appearances in light of his own conclusion that the democratic evolution of capitalism into socialism is quite feasible.

IDEOLOGY OF THE "OBJECTIVE INTERESTS" OF THE WORKING CLASS

The Bolshevik ideology, like any other, has two dimensions: negative and positive. Its negative side consists in simply rejecting bourgeois democracy as formal, apparent, and even illusory. There is, of course, no easy denial of the obvious, i.e., that through the electoral process citizens freely express their preferences. What the Bolsheviks claim, however, is that citizens most often are not aware of their real, objective interests because of indoctrination, manipulation, lying, their low level of education, and the like.

The distinction between subjective and objective interest has played a major role in the ideologizing Marxism. Such distinction belongs to a group of Marxist categories designed originally to transcend subjectivism. Let me mention here only the contrast between the *subjective and objective meaning* of human acts, and that between *subjective and objective responsibility* for them. In my book *In Search of Democracy in Socialism*,[5] I analyzed step by step the transformation of "objective meaning" and "objective responsibility" into typically ideological notions.

---

5. (Buffalo, N.Y.: Prometheus Books, 1981).

Let us, however, return to subjective and objective interest by quoting a theorist who successfully uses this distinction in his critique of bourgeois democracy. Here is what Isaak D. Balbus means by "objective interest": "Interest in this sense, then, is objective because it refers to an effect by something on the individual which can be observed and measured by standards external to the individual's consciousness."[6] Balbus then effectively applies the concept of "objective interest" to social groups and classes.

I would like to point out, however, that one still ought to be very careful because an important difference between the objective interest of an individual and that of a social group or class should be kept in mind. The former is limited to the life span of individuals and their immediate offspring, whereas the latter might refer to numerous generations. We are sick and tired of the manipulative ideology of sacrificing the subjective interest of the working class in the name of its objective interests. Still, there is no doubt that one can meaningfully speak about sacrificing a generation's interests for the sake of future generations. It should, however, be freely decided by the affected generation and not superimposed on it, allegedly in its own interest, by a self-styled vanguard.

As is obvious, I also think that subjective and objective interest can in principle be distinguished. The assumption that everybody always knows his or her real interest best is simply wrong. That is why we turn to experts and other advisers— of course, under the condition that we keep the final decision in our own hands. In collective life the crucial question is what is deduced from the distinction between subjective and objective interest: the need for less or more public interaction and discussion, less or more access to mass media, freer elections,

---

6. "Concept of Interest in Pluralist and Marxist Analysis," *Politics and Society* 1, no. 2 (February, 1971): 151-177.

or, on the contrary, the need to extinguish it all in the name of the knowledge monopoly of a vanguard.

Marx said almost nothing about the transformation of the proletariat's objective into subjective interests, i.e., the development of its class consciousness. He also wrote nothing persuasive about how to avoid the danger of a self-appointed interpreter and representative of the objective interests of the working class.

Thus we have come in our analysis to the *positive* dimension of the Bolshevik ideology. This ideology emerged historically by referring to the objective-historical interest of the working class. The Bolshevik ideology proceeds from the distinction between the interests of which the working class is conscious ("subjective") and the interests of which it is not conscious ("objective"). Unlike it, bourgeois-democratic ideology does not involve any difference between preferences and interests: what voters express at elections as their interests is considered to reflect their interests.

Since the Communist party, as the alleged vanguard of the working class, represents first and foremost these "deeper," "real" interests,[7] there is no need for forms of political life and organization by which that class would freely express and realize its preferences. The political vanguard claims to represent objectively the interest of that class, if necessary even against the expressed will of the workers. Bolshevism was quickly transformed into an ideology of statism. In statism eventual *market* preferences of the working class are suppressed through *politocratic planning* resting allegedly on the objective interests of that class. Having introduced such a degree of

---

7. It is interesting that the de facto division of the party into a supervanguard (leadership) and vanguard (membership) has never been justified by invoking the objective interests of its members allegedly interpreted and represented by the leadership—but only by referring to the ideologized principle of "democratic centralism."

arbitrariness, it was easy to transform "*objective interests*" into "*historical interest*" and finally even into a "*historical mission*" of the working class. In this way, a cognitively useful concept became a metaphysical-ideological one, and alleged objectivism turned out to be subjectivism (of the Communist party).

This ideology has a built-in, in-depth defense. Although admittedly ("on the empirical level") there are no real opportunities for the working class to express its political preferences, it is said to be in *substance and essence* (I would say metaphysically) objectively represented by its vanguard, which even calls the proletariat the ruling class. It is symptomatic enough, however, that in this jargon a formulation like "working class in the metaphysical sense" is unknown, although on the other hand the expression "working class in the empirical sense" is very often used.

Anyone who wants to bring into question such an ideology must first subject to criticism this notion of objective interest. In ideologized Marxism this interest has been turned into a metaphysical category. For me, in this context, metaphysics consists of principally *unfalsifiable* statements.

Bolshevik ideologists never established *when and under what conditions* it could be verified in practice whether their party truly represents the interests of the working class. They never promised to allow the working class itself, within a precisely defined period, to express its own interests. Neither Stalin nor Khruschev after him predicted when workers would be enabled to express their interests in free elections on "objective representation," although both were want to say exactly when the USSR would complete the construction of socialism and move on to building communism.

A contemporary critique of ideology is impossible without a fundamental critique of its language. Ideological terminology belongs, so to speak, to the sphere of the "objective spirit": this is how ideology is reinforced and transmitted, often with meanings that have little to do with the intentions of those who initiated that terminology. Ideological distortion often occurs in the very assigning of names. From the very first move, in the choice of words, problems are largely minimized or completely suppressed. It is those who have the power to assign names to social phenomena that dominate the world ideologically. Their power is augmented by the naive linguistic realism of the broad masses of people who are incapable of entirely separating words from the social phenomena they signify. Here I shall analyze three examples from the history of bolshevism: "dictatorship of the proletariat," "democratic centralism," and "transition period."

What would a terminology undistorted by ideology look like? Instead of saying "democratic centralism," we could say, for instance, *revolutionary centralism:* there is a revolutionary organization that operates from the underground and cannot survive in the face of the repressive state apparatus unless it is centralistically organized. Why not simply call it revolutionary centralism: revolutionary in the descriptive, not in the evaluative sense? Such centralism under the described conditions can truly constitute the only realistic mode of organizing revolutionaries.

Furthermore, there are professional revolutionaries who take power and establish a dictatorship. But, it is an empirical question as to who constitutes the social base of that dictatorship: the proletariat, the peasantry, or perhaps these revolutionaries stand, above all, for their own interests, objectives, and conceptions. In other words, this question should not be closed by the choice of terminology, and it is better, therefore, that we simply talk about *revolutionary* (not proletarian) *dictatorship.*

Or take, for instance, the notion of the "transition period." This period defines the temporal and qualitative framework of both "proletarian dictatorship" and "democratic centralism." It plays an essential role in the in-depth ideological defense of bolshevism: whatever a ruling communist party does has to be construed and justified as a set of unavoidable means during the "transition period." Thus a society and its ruling political organization are not seen as they are, but rather as moments of a super-optimistic, super-deterministic, and super-teleological historical scheme. But what would be an alternative? Well, simply to call it the *post-revolutionary period.*

Unfortunately, all important questions are closed by the very choice of language. "Proletarian dictatorship": since the dictatorship is proletarian in its very name, one cannot meaningfully pose the question of the true relationship between the proletariat and revolutionaries in power. (E.g., do the latter really represent the proletariat, or have they become alienated? How are they to be controlled?) Furthermore, when one says "democratic centralism" then one cannot quite properly raise the question of its democratization, since it is democratic per *definitionem.* Finally, the "transition period": since this notion in itself says almost nothing (every period in history is a transition between two periods), we have "only" to pose a question regarding the nature of the condition into which we are *really* passing. But just this question has already been answered in an *a priori* fashion.

THE TRANSPARENCY OF STATISM

It is the job of science, according to Marx, to *penetrate* "appearance" (*Erscheinung* and *Schein*) and "form" and reach the "essence," "reality," and "content." That is why R. Heilbroner calls Marx's science "socioanalysis."

Here is how G. A. Cohen describes Marx's view: "Illusion is constitutive of class societies. I say 'constitutive' because more is claimed than that the members of a class society acquire false beliefs about it. The falsehood maintains its grip by permeating the world they experience: their perceptions are false because what they perceive is a distortion of reality.

According to Plato (on some interpretations) men observing the material world are under illusions not because their thoughts fail to correspond to it, but because they faithfully reflect what is an illusory world. Marx was not untouched by the philosophical tradition which Plato began; thus when he writes that workers take seriously the appearance (*Schein*) that their labor is fully rewarded, the phrasing shows that he thinks of the appearance as an attribute of the reality. It is only derivatively a reflection of reality in men's minds."[8] Furthermore: "Things do not seem different to a worker who knows Marxism. He knows they are different from what they continue to seem to be. A man who can explain mirages does not cease to see them."[9]

Ludovico Silva writes impressively about "Marx's infinite power to *invert* perspective, to uncover the reverse side of social phenomena which economists, philosophers and politicians observe only from the front side, seeing only their character as it appears."[10] He goes on to say: "The perceptual or stylistic module of transmitting this 'heightened consciousness' stems from Marx's developed ability to build sentences and periods, where in the ascending phrase he ironically presents the front side of the phenomenon ('The peasants' mortgage on heavenly goods') and in the descending phase he reveals the reverse side, the actual structure hidden behind the appearance ('constitutes a guarantee of the bourgeois mortgage on the peasants' goods')."[11]

Considerable discrepancy between apparent forms and essence can only be explained by the *structural* complexity and separation of spheres of social life in capitalism. Here I

---

8. See G. A. Cohen, *Karl Marx's Theory of History. A Defense* (Princeton, N.J.: Princeton University Press, 1978), pp. 330-331.

9. Ibid.

10. *Marx's Literary Style*, Serbo-Croat translation from the Italian, Belgrade, p. 99.

11. Ibid.

mean, above all, the separation of "civil society" from the state (somewhat simplistically: economy from politics). But separation also exists in the economy: that is why Marx treats the sphere of circulation as an appearance behind which lies production as the essence.

Marx's type of critique of ideology was developed for a social formation with *economic* dominance. Unlike capitalism, statism ("socialism") constitutes a social totality with *political* dominance. Under capitalism, politics *structurally screens* those dimensions of power that result from the economy, which is why critique should penetrate these political "forms" and "appearances" (*Erscheinung* and *Schein*) to the "content," "essence," and "reality." Such contrasting of politics and economy is not possible in a critique of the ideology of statism, because here virtually all power, including economic power, stems form politics. In other words, Marx's paradigm for a critique of ideology is not applicable to statism.

To be sure, appearance-form and essence do not "directly overlap" (Marx's expression) in statism as well. However, in capitalism the appearance-form *misleads* in terms of essence, whereas in statism it *leads* to the essence. A scholar giving an exact description of statist forms does not deal with illusions: he takes a direct and important step toward presenting the essence of statism. For example, elections where as a rule there are no real counter-candidates show the essence of the statist system: monopoly over politico-economic power. Those who think that this is a question of the *same* "forms" as in democratic capitalism are wrong. Election campaigns without opposition candidates, elections whose outcomes are predetermined, parliaments and trade unions as transmission belts of the ruling party—these are all *empty* democratic forms, procedures, and institutions, taken to surrealist dimensions.

That is why statists in their society can hardly tolerate even the "purely" descriptive sociology of power. Sociologists must not simply enumerate leaders and define the province of their power, without, at the same time, claiming that these power-holders are merely vanguard representatives of the objective and historical interests of the working class and "all the people." But, the ideological character of this notion of "representation" is obvious, because those who are ostensibly represented never have the opportunity of actually electing their representatives.

Needless to say, knowledge of the essence depends on the kind of theory we have at hand: the more precise and deeper the theory, the closer we are to the essence of higher order. For instance, if we set out from the Marxist conception that the "ruling class" must *own* the means of production, we shall be unable to identify statists as a ruling *class*. They obviously are no *private-ownership* class. Furthermore, the notion of *state ownership* only obscures the true nature of this group and its power.[12] Since they are in no way the owners of the means of production, not even owners of a *collective* character, it is alleged that all citizens are collective owners ("state ownership" = "social ownership") of the means of production in statism. But how can there be *social ownership* without democracy? In actual fact, the ruling statist group holds a *monopoly of control over the state and, through it, over the means of production.* What other kind and degree of power would a group have to have before we could designate it as the ruling *class*?

We have established that Marxism still has a substantial potential for a critique of capitalist ideology. On the other hand, it has also been used to conceal the real nature of statist power. If, for instance, surplus value can by *definition* only be appropriated by the *owners* of the means of production, then Marxism inadvertently helps the ideology that says that there

---

12. It obscures it so much that it seems to many that only one "small step" separates statism from socialism, the step being socialization of "state ownership," which is allegedly already one form (albeit a lower one) of social ownership.

is no exploitation in "socialist" statism. Further, if *homo duplex* (the abstract citizen versus the egoistic individual) emerges only because of market competition, then it follows that "utilitarian" human relationships reign only in capitalism, not in statism. The point is, however, that in comparison to capitalism, in statism the main source of "utilitarianism" lies in politics, not in economics.

In statism there is no "invisible hand," to use for our purposes a term from a different context and with a different meaning. Contrary to statism, the capitalist economy conceals a number of power holders whose power is difficult to see in the political sphere. Democratic-capitalistic ideology tries to present the political power structure as entirely transparent: no hidden major limitations lie "behind" democracy. Statist ideology, on the other hand, suggests that "Marxism" opens up one's eyes to what is invisible to non-Marxists: the Communist party allegedly always stands for the "objective" interests" of the working class.

The description of power in one sphere of statism reveals the politiocratic essence of the entire power structure. Russian dissident and emigre writer A. Zinoviev bases his literary procedure on just that: without any explicit ideological exposure, he "merely" describes the everyday life of the intelligentsia in the USSR, because he knows that the description of power in one field says enough about the nature of global power.

Marx forecast that the *classless* society, unlike capitalism, would be transparent. It is a great irony of the history of Marxism that a new class society, which in many ways is much more transparent than capitalism, has been created in the name of Marxism. And this leads to yet another irony of history. G. A. Cohen writes: "If we conjoin Marx's conceptions of socialism and science, we obtain the conclusion that socialism renders social science superfluous. It has to function in a world that has abolished the discrepancy between the surface of things and their true character."[13] In fact, it is in statism, not socialism, that genuine social science has largely become "superfluous" because of ideology.

Many will be surprised by my contention that statism is much more transparent than capitalism. How is that possible when in the most progressive capitalism there is democratic public opinion and freedom of social science, whereas in statism virtually everything regarding power is a state secret? But I have been talking only about the *locus of power*. In capitalism it is much easier for domination over the state to conceal itself, because it stems from the economy and not politics. In statism, on the other hand, political monopoly of power in all spheres constantly strikes the eye. In capitalism, one still has to *penetrate* the political-legal sphere and ideological illusions that emerge if we confine ourselves to it. In statism, however, everybody knows that political power is the focal point and source of power. What can be concealed is not the fact that power is concentrated in the so-called *nomenklatura*, but that it constitutes a new ruling class, its exact functions, what the relationships are between groups and individuals within it, how it lives in private, what privileges it enjoys, and much more.

G. A. Cohen draws the following comparison between feudalism and capitalism in terms of transparency:[14]

	FEUDALISM	CAPITALISM
That surplus product is extracted	is evident	is concealed
That human relations are "utilitarian"	is concealed	is evident

---

13. *Karl Marx's Theory of History*, p. 336f.
14. Ibid., p. 333.

What should be added to this table is domination over the state as one of the indicators, and the entire comparison should be extended to include contemporary statism:

	FEUDALISM	CAPITALISM	STATISM
That surplus product is extracted	is evident	is concealed	is concealed
That human relations are "utilitarian"	is concealed	is evident	is concealed
That a group dominates the state	is evident	is concealed	is evident

6

From Bolshevism to the Ideology of "Real Socialism"

THE BOLSHEVIK IDEAL-LOGICAL PARADIGM

Communist parties have inherited from Lenin and other great Bolsheviks and ideal-logical[1] paradigm. In terms of this paradigm, the Bolsheviks understood themselves and the world, which they tried to disqualify ideologically and to change through revolutionary activity.

Apart from the ideal of a communist society, the following ideas exercised key influence on the Bolshevik's (self)understanding: proletarian revolution; a party of (professional) revolutionaries organized in a democratic-centralist manner; dictatorship of the proletariat; the party as the representative of the objective and historical interests of the proletariat; the transition period between capitalism and communism.

As *Marxists* the Bolsheviks faced the problem of how to

---

1. In chapter 4, the section titled "What Should Ideal-logy Be?" I explained what this term means.

explain to themselves and to others the possibility of socialist revolution in backward Russia. The idea of a centralized party of revolutionaries did not suffice; they needed a radical revision of the Marxist conception of revolution. This revision transformed a revolutionary (philosophical-social) theory into a revolutionary ideal-logy. Bolshevik ideal-logues believed that Marx's goals could also be achieved under radically changed conditions and through radically changed means.[2] In connection with it, one could almost say—ironically—that revolution sets only those goals it cannot achieve.

Two ideas were of key importance for the Bolshevik ideal-logical revision of Marx's conception of revolution: the "weakest link" and the "permanent revolution." The idea was rejected that capitalism must first be broken in the areas where it was strongest; instead an opposite starting point was adopted. However, this idea in itself could have bolstered expectations of a bourgeois-democratic, but not of a socialist, revolution in backward Russia. Hence the need for the idea of permanent revolution. The bourgeoisie in Russia was said to be too weak and afraid to lead a bourgeois-democratic revolution with any consistency, and so the working class (led by its vanguard) had to be the main vehicle of this revolution. Once the working class carried out the revolution, there was no need to give the liberal bourgeoisie a leading role—on the contrary, the next, socialist stage of revolution had to be embarked on immediately.

The international aspect of the idea of permanent revolution lay in the expectation of a world, or at least a West European, revolution. This is the link with classical Marxism: There is

---

2. Here is a further example of ideal-logical (self)delusion. When the relation of forces in international communism radically changed, some time ago, the slogan of "different roads to socialism" was adopted. But, when one takes a close look at the nature of the means, methods, and conditions this implies, it becomes clear that the goal cannot be the same "socialism," but rather, different "socialisms."

no hope for the proletarian revolution in backward Russia unless revolution succeeds at least in capitalist Western Europe.

Neither the concept of the weakest link nor that of permanent revolution were seriously problematized. In reality, if a country constitutes the weakest link of capitalism and imperialism, at the same time it can be the weakest link in the hands of revolutionaries as well, thus offering the least chance for building a new socialist society.

With respect to the permanent revolution, we have to ask ourselves: How did the assumption ever come about that the working class, as an *economic* class (one even weaker than the bourgeoisie in Russia) could lead a bourgeois-democratic revolution and be powerful enough to control post-revolutionary *political* development? Moreover, how could it be assumed that such a class could introduce and carry out a dictatorship?

The victory of Stalinism is usually explained by the overall weakness of the Russian working class, which was virtually decimated in the civil war and therefore unable to dominate the post-revolutionary process. The problem is much deeper than that, however. What were the grounds for presuming that a small and weak working class would be capable of safeguarding for itself *bourgeois-democratic* revolutionary achievements, not to mention its domination over the new state apparatus and the safeguarding of Soviet achievements? After the October Revolution, not only was the *Soviet* dimension gradually eliminated, but the achievements of bourgeois-democratic revolution were also destroyed, such as freedom of assembly, association, parties, and independent trade unions. After February 1917, the working class clearly had more to lose than "its chains."

In my opinion, at the heart of the idea of permanent revolution lay a mistaken dichotomy stemming from classical Marxism: *capitalism or socialism*. History has shown that at the very least a trichotomy should have been taken into account, because what emerged was the *statization* of the revolution (culminating with Stalinization).

As for the international aspect of permanent revolution, mention is usually made of Lenin's and Trotsky's mistakes in predicting a European revolution. However, it is usually forgotten that their actions often alienated potential supporters in the West. When they realized that their revolution lacked the inner strength to maintain itself, why did they narrow its social and political base by eliminating all other socialist parties and organizations and thereby alienate the democratic West, and a good part of the workers' movement there?

I come back to my claim that the idea of permanent revolution was not seriously problematized, so that the assumptions about the very first stage were unfounded. Now, let me take this thesis a step further: The very idea of *two stages* was unfounded and proved to be more harmful than useful. This holds true even more for the assumption that the second stage was bound to have a socialist character.

True, behind the idea of permanent revolution there was correct presentiment that Russia was entering a long period of revolutionary upheaval. Indeed, this is what happened: a revolution in 1905, two revolutions in 1917, and a third revolutionary upheaval that started in 1928–29 and was the begininng of the final victory of Stalinism by means of complete statization and terrorist collectivization.

In theory-ideology, however, on the basis of a philosophical-historical scheme, a conclusion was drawn on the lower and higher stages of revolution, instead of on the *long-term revolutionary*[3] *process with different, even contrary tendencies: capitalist, statist, and socialist.* The statist possibility and tendency was not seriously considered, because of the theoretical

---

3. In contrast to many Marxists, I use "revolution" as an empirical historical notion, not as a category of the philosophy where revolution is linked to only progressive changes.

dichotomy: either capitalism or socialism. Here one should again raise doubts about the claim of the October Revolution's unquestionably socialist character. In its unequivocally anti-feudal and anti-capitalist orientation, this revolution, from the beginning, developed a tension between its socialist (Soviet) and statist components.

This was an *increasingly inferior and envious communism* that, faced with a far more advanced capitalism, developed a variety of arrogant ideological appraisals, compensations, and rationalizations. That is why the findings of political psychology should be applied to its examination. Such a study should start by examining the Bolshevik perception of the threat of "capitalist encirclement" and their slogan of "catching up and overtaking" the capitalist world.

THE IDEAL-LOGICAL AND PRAXOLOGICAL CRITERION

I call classical Bolshevik ideal-logy *socialist realism* because the assumed *tendency* of movement toward communism was, for it, more important for determining the character of social reality than any facts. I have borrowed the name from the official Soviet literature and culture (where the main demand is to "reflect" this tendency), but I expanded its range of application to the entire Bolshevik ideology.

It may be an exaggeration to say, like some moral philosophers, that man is bound by *nothing* moral if it is only the future that binds him. But it would be hard to deny that *little* binds us morally, if only the future binds us. This is the old problem of the relationship between the ethics of ideals and the ethics of means.

Leon Trotsky's defense of bolshevism in the 1938 essay "Their Morals and Ours" bears all the traits of ideal-logy. Twenty years after the revolution, he still defended all the Bolshevik actions as unavoidable means to achieving communist goals along the lines of "the end justifies the means." He staunchly refused to see in any of these acts even the possible germ, let alone the fertile soil, for the victory of Stalinism. This is as though Stalinism was nothing more than the abuse and betrayal of bolshevism, and not one of its tendencies and currents as well.

Writings on negative utopia give a striking picture of how Stalinism functions at the peak of its power. It would be even more interesting, in my opinion, and more important from the practical point of view, to concentrate less on the ultimate result and more on the process that led up to it. This is the theme of the *emergence of negative utopia*. There is no doubt that Lenin, Trotsky, and other leading Bolsheviks bear a (co-)responsibility in this respect. It was under their leadership that the *structural possibilities and tendencies* were created that led in this direction. Who else, if not Lenin, bears responsibility for the extreme formulations on the "dictatorship of the proletariat," as expressed in the following: "a special form of cudgel, *rien de plus*."[4]

This ideology did not always reject unpleasant truths. They were "merely" proclaimed "small" and "partial," truths as compared to a "big," "epoch-making," and "whole" truth. The "historical tendency" of movement toward communism carried greater weight and reality than "individual facts taken in isolation from this tendency."

While ideal-logy primarily cites proclaimed historical goals, praxological critique primarily establishes the actual consequences that ensue in the said struggle to realize these goals. Bolshevik ideology has bountifully applied the principle of "objective meaning" (all consequences) to everybody else, while assessing itself primarily in terms of intended, proclaimed consequences ("subjective meaning"). A theoretical obstacle to perceiving the

---

4. V. I. Lenin, *The Lenin Anthology*, edited by Robert C. Tucker (New York: Norton, 1975), p. 490.

ideal-logical character of the Bolshevik approach to the future was also Marx's conviction that he himself had not set up any ideal (of communism), but only scientifically described the necessary tendency of social development (toward communism).

Lenin's *State and Revolution* is a manifesto of revolutionary ideal-logy. Its example can be used to study the relationship between the most general political program, the program of action, and practice. What in the most general program actually obligates revolutionaries, and what, on the other hand, constitutes their ideal-logical (self)delusion, does not become quite clear until they acquire power. No wonder skeptics say that virtue flowers where there is a lack of temptation.

Even worse are some of Lenin's and Trotsky's actual instructions regarding terror. Would their instructions have been quite the same if the following rule had been one of the minimal principles of revolutionary morals: Anyone who orders something to be done that clearly clashes with the humanist legacy is duty-bound to set a personal example by being the first to take such action. But, in an atmosphere marked by "democratic centralism" and "revolutionary discipline," a particular kind of *homo duplex* emerged: leaders formulate ideology, determine programs, and issue directives, while others are, in their eyes, suitable for doing the "dirty work."

In speaking about the (co-)responsibility of leading Bolsheviks for the victory of the Stalinist tendency, it would be unjust to overlook the West's own (co-)responsibility. It has long since been established that with their extremely revengeful attitude toward Germany after World War I, leaders in the West had an unintentional part in creating and strengthening the chances for Hitlerites to come to power. The same type of responsibility can be applied to Soviet history in the twenties and thirties: first counter-revolutionary foreign intervention, then isolation of the USSR, and finally pushing Hitler and Mussolini against the USSR—all such steps played into the hands of Stalin and the worst Stalinists.

Of course, the furthest thing from my mind is to equate Stalinism with other currents and tendencies of bolshevism. Those who limit comparisons between the period before and after 1928 to a difference of *degree*, and deny the difference in *kind*, would do well to ponder from what standpoint they are doing this. Only an *abstract theory deprived of concrete moral and generally humane sensitivity* is able to relativize to this extent the difference between Stalinism on the one hand and Leninism and Trotskyism on the other. Stalinists committed millions of state crimes. The Twentieth Congress of the Soviet Communist party was not enough for a massive catharsis inside and outside the USSR. What is needed is a symbolic trial of such great state criminals as Stalin and his henchmen.

Because of its "Marxist" genealogy, Stalinism is an ideology with immense intellectual ambitions. Indeed, Stalinism's super-ideological character is unparalleled: It depicts itself as a scientific ideology, and all others as unscientific. Some critics seem to accept Stalinist self-understanding and talk about its "scientism" and "positivism." True, there are elements of both in Stalinist self-understanding, but no less true is the fact that Stalinist "science" cannot intellectually stand up to even the simplest of positivistic criteria and analyses. Otherwise how could one explain the great *intellectual effectiveness* of a kind of "neopositivist Marxism" in reckoning with Stalinist ideology?

During Marxism's Stalinization, even Marx's way of criticizing ideology assumed an ideological character. As a rule, Marx first endeavored to show that a given world view really presented a distorted or false picture of the world, and only then asked why this was so and looked for the answer in the place and interests of a group or class in the social division of labor.

It is a mistake to believe that Stalinists only change the order of the moves, so that the social root of the criticized ideas comes in first place, while the question of their truthfulness comes second. Stalinists actually do not examine whether someone's ideas are true or not, because they are concerned with discrediting such ideas at any price. The position they stick to is that the presumed social origin of such ideas in itself implies their untruthfulness, which, of course, is a generic fallacy. Still worse, the social roots of ideas are not examined but established *a priori*. Ideas that differ from Stalinist ideas have *by definition* an undesirable social origin. Thus truthfulness or untruthfulness of world views is deduced from the *transcendental* connection between social groups and their alleged progressiveness or regressiveness. However, it is actually the other way round: The degree of concern for the truth as a social value is one of the criteria of progressiveness of social groups.

FROM "SOCIALIST REALISM" TO "REAL SOCIALISM"

By the end of the 1960s the ruling ideology in the USSR and Eastern Europe, burdened on the one hand by the Stalinist legacy and challenged on the other by reform communism, was transformed explicitly from the former ideal-logy to a kind of real-logy. In order to justify itself ideologically the system could not, as it had done previously during the period of "socialist realism," continue to invoke mainly a communist utopia. Rather, the system began to invoke primarily the fact that it was a reality (of course, "reality," as it was defined by ideology, and not as it actually was). Hence the new official name: "real socialism." This new formula suggested that any alternative "socialism with a human face" was pure utopia.

In the 1950s and 1960s the adherents of the communist ideal-logy reacted in different ways to the Stalinist legacy. Some readily sacrificed their communist ideals, others became disillusioned and passive, while yet others revolted against Stalinism precisely in the name of their communist ideals. Milan Kundera excellently describes this last group in his *Book of Laughter and Forgetting:*

Yes, say what you will the Communists were more intelligent. They had a grandiose program, a plan for a brand-new world in which everyone would find his place. The Communists' opponents had no great dream; all they had was a few moral principles, stale and lifeless, to patch up the tattered trousers of the established order. So of course the grandiose enthusiasts won out over the cautious compromisers and lost no time turning their dream into reality: the creation of an idyll of justice for all. . . . And suddenly these young, intelligent radicals had the strange feeling of having sent something into the world, a deed of their own making, which had taken on a life of its own, lost all resemblance to the original idea, and totally ignored the originators of the idea. So those young, intelligent radicals started shouting to their deed, calling it back, scolding it, chasing it, hunting it down. If I were to write a novel about that generation of talented radical thinkers, I would call it *Stalking a Lost Deed* . . . .

Historical events usually imitate another without much talent, but in Czechoslovakia, as I see it, history staged an unprecedented experiment. Instead of the standard pattern of one group of people (a class, a nation) rising up against another, all the people (an entire generation) revolted against their own youth. Their goal was to recapture and tame the deed they had created, and they almost succeeded. All through the 1960s they gained in influence, and by the beginning of 1968 their influence was virtually complete.[5]

The *alienation* of the communist idealists' deed and their effort to *de-alienate* it would constitute an excellent but separate topic. Here, I would like to mention only that the official ideology of "real socialism" represents in good part a response to the communist idealists' challenge in Eastern Europe.

Whenever the ideology of "real socialism" came into conflict with reformist endeavors it, manifested one of its hidden faces: cynical consciousness. Any movement for democratic-humanist

---

5. (London: Penguin Books), 1981, pp. 8-14.

socialism was suppressed by *force*, but the ruling ideologues continued triumphing by claiming that the "existing socialism" represents the single *realistic* possibility.

Unlike the other countries of the socialist camp, the Soviet ideology, prior to Gorbachev, came forward in the name of both "real" and "developed" (or "mature") socialism. In other words, there existed an international ("internationalistic") hierarchy of "real socialisms." That is why Edward Gierek proclaimed "developed socialism" as the goal of Poland. "Real socialism" in Eastern Europe was to "catch up" with "developed socialism" in the USSR, although this in turn had not yet "caught up" with the most developed capitalism.

But if socialism is already developed, why is the transition to communism relegated to the indefinite future?[6] The latest congress of the Communist Party of the Soviet Union had to opt for an indefinite period of the further development of socialism. On this topic Mikhail Gorbachev said:

> While some suggest that references to developed socialism should be completely removed from the Program, others, on the contrary, believe that this should be dealt with at greater length. The draft sets forth a well-balanced and realistic position on this issue. The main conclusions about modern socialist society confirm that *our country has entered the stage of developed socialism.* We also show understanding for the *task of building developed socialism set down in the programme documents of fraternal parties in the socialist countries.* At the same time, it is proper to recall that the thesis on developed socialism has gained currency in our country as *a reaction to the simplistic ideas about the ways and terms of carrying out the tasks of communist*

---

6. During Khrushchev's rule in the Program of the Communist Party of the Soviet Union (1961) it was declared that the "dictatorship of the proletariat had fulfilled its historic mission" and was transformed into a "state of all people." Brezhnev included the formula in the Constitution of the USSR (1977).

construction. Subsequently, however, the accents in the interpretation of developed socialism were gradually shifted. Things
were not infrequently reduced to just registering successes, while
many of the urgent problems related to the conversion of the
economy to intensification, to raising labor productivity, improving supplies to the population, and overcoming negative
things were not given due attention. . . . Today, when the Party
has proclaimed and is pursuing the *policy of accelerating socio-*
economic development, this approach has become unacceptable. . . . As for the chronological limits in which the *Program*
targets are to be attained, they do not seem to be needed. The
faults of the past are a lesson for us. The only thing we can
say definitely today is that the fulfillment of the present Program
goes beyond the end of the present century.[7]

Critics of the ideology of "real socialism" have already pointed
to its tautological nature: it seeks to justify reality by invoking
"reality." According to the critics it is also a conservative
ideology: "reality" is proclaimed the norm.

T. H. Rigby has noted, correctly, that Max Weber's typology
of legitimation (traditional, charismatic, and formal-legal) is
incapable of encompassing the Soviet type of social order:

The predominant orientation of these command-structures is
towards *goal achievements*, rather than towards the application
of rules, which Weber correctly identifies as the predominant
orientation of the public bureaucracies of Western "capitalist"
societies. . . . Consonant with this, the legitimacy claimed for the
commands issuing from this system and for those holding office
under it is framed in terms of "*goal rationality*" rather than the
formal-legal rationality of Western "capitalist" systems. . .
Though some essentially rule-applying bureaucracies are present,
the predominant bureaucratic mode is the *task-achieving* mode.
Accordingly the central role in the political system is played by

---

7. *Political Report of the CPSU Central Committee for the 27th*
Party Congress (Moscow: Novosti Press, 1986). (Italics mine.)

institutions concerned with formulating the goals and tasks of the constituent units of society and supervising their execution. Consonant with this, the legitimating claims of the political system of those holding office under it, and of the latters' commands, are validated *in terms of the final goal* ("communism") from which the partial and intermediate goals set by the leadership are allegedly derived and to which individual goals should be subordinated.[8]

Rigby is right when he says that the Soviet system tries to legitimize itself through "goal" rather than "formal-legal" rationality. However, it is not true that also under Brezhnev this ideology invoked mainly the *"final goal" (communism)*. Rather, the ideology's ambition became more moderate: further development of "real" and "developed" socialism. The category of "goal rationality" can mislead us if we do not clearly distinguish between "socialist realism" and "real socialism."

After all, seventy years after "Marxism-Leninism" came to power, there are very few who are willing to forgive its faults because of its alleged "historical tendency to move toward communism." As soon as "socialism" begins ideologically to rely mainly on its own "reality," it is inevitably judged on the basis of its *performance*. And it is exactly at this point that the troubles begin, ranging in intensity from stagnation to the obvious crisis in some countries. "Real socialism" remains functionally inferior to developed democratic capitalism. Even the social policies of "real socialism" do not always compare favorably with those of the social-democratic "welfare states" in the West, especially those in Scandinavia.

The following question presents itself as well: What kind of "real" socialism is this when, for instance, in Poland a workers' movement of many millions revolt against it? Those "patriots"

---

8. T. H. Rigby and F. Feher (eds.), *Political Legitimation in Communist States* (New York: St. Martin's Press, 1982), introduction by T. H. Rigby, pp. 10–20. (Italics mine.)

who introduced martial law against this movement suggested *implicitly* that the Polish people should accept them as representing a *lesser evil* than eventual direct foreign occupation. Since a radical alteration of the power structure in Eastern Europe is not possible without risking universal nuclear destruction (*absolute evil*), the existing evil appears *relative.* This is a good example of the role of evil in ideological justification. According to Werner Becker, we have in this regard reached the limits of the customary understanding of legitimation:

> One should not ignore the fact that in the second half of this century we have perhaps reached the limits of the classical understanding of legitimation. This understanding has never before in history envisaged the possibility that the state rulers might become *almighty* in carrying out their will. . . . In these states [of the Eastern Bloc] the power structure of "real socialism" is being preserved only through the threat that the Soviet Union would begin the big war in case of states' instability in its sphere of influence.[9]

In Poland the official ideology has, in my view, entered the third phase: from *distorted* through *false* to *mendacious* consciousness. All Polish people reject this mendacious ideology and even the officials do not believe in it. Why, then, do they employ it? Here is one of the suggested explanations:

> This propaganda does not seek to persuade anybody, it is aimed to defeat. It says: look, listen, we can say whatever we wish, any lie, we can spread dirt on whomever we choose, we can offend, humiliate, provoke you, we can attack anything which is dear to you, we can sentence the finest patriots for treason and decorate traitors with medals—we can do anything, and you can only sit in front of the television and clench your fists. Listen and be

---

9. *Die Freiheit, die wir meinen,* R. Pipper, München, 1982, p. 17f.

silent. This is how strong we are. This is our revenge for your attempt to dream about a better world.[10]

But the problem is more complex. That which is possible in Czechoslovakia is, for instance, no longer possible in Poland. It is true that the Polish statist class would also like to monopolize the public sphere, excluding from it all alternative languages. But it no longer possesses this kind of power. This class, it is true, still uses "Marxist-Leninist" formulae, but it does not expect to represent successfully untruths as truths.

Rather, we are dealing with the power-holders' signals to one another, and even more to the Soviet Union: They will not permit a fundamental alteration of the constellation of forces at any price. If they were to abandon the official ideological language, the USSR would come to the conclusion that its vital geo-strategic interests are imperilled. And the system of the statist "nomenklatura" still has an international dimension as well, although in some countries this external control of cadres has more of a negative and indirect character than a positive and direct one.

However, the mendacious consciousness has not prevailed for the first time just in Poland. During post-1968 "normalization" in Czechoslovakia, the majority of those seeking to retain their positions or jobs had to reject publicly the "Prague Spring" and to accept publicly "fraternal help." The power-holders were not interested in the least whether this was done sincerely or not. Here is another interesting contribution to the analysis of the reality and nonreality of Czechoslovakian "socialism":

> The manager of a fruit and vegetable shop places in his window, among the onions and carrots, the slogan: "Workers of the World, Unite!" Why does he do it? . . . Let us take note: if the greengrocer

---

10. Jacek Federowicz, "Let Us Have the Censor," *Index on Censorship* (October 5, 1985).

had been instructed to display the slogan, "I am afraid and there-fore unquestionably obedient," he would not be nearly as indif-ferent to its semantics, even though the statement would reflect the truth. The greengrocer would be embarrassed and ashamed to put such an unequivocal statement of his own degradation in the shop window, and quite naturally so, for he is a human being and thus has a sense of his own dignity. . . . Thus the sign helps the greengrocer to conceal from himself the low foundations of his obedience, at the same time concealing the low foundations of power. It hides them behind the facade of something high. And that something is *ideology*.[11]

However, those who expected that the statist system would perish the moment it entered the ideological impasse of menda-cious consciousness were mistaken. Even ideologically defunct systems can continue to exist, especially if the international constellation of forces is favorable to them. Besides, if the official ideology of "Marxism-Leninism" loses influence, it does not mean that the statist system cannot have a different kind of ideological support. As an example we have already mentioned the ideology of the "lesser evil." True, this ideology in itself does not have enough strength to make possible a return from *mendacious* to *distorted* consciousness.

Reliance on the victory in World War II, superpower status, and, generally, the patriotism of soviet citizens are usually treated in the literature as a "secondary" ideology. But why not "primary," since this ideology *de facto* is no less important for the system's legitimation than "Marxism-Leninism"? After all, in the USSR "Soviet patriotism" is officially regarded as a component of "Marxism-Leninism" and not as something separate from it. Moreover, precisely those generations for whom the experience of World War II remains decisive still

---

11. Vaclav Havel, "The Power of the Powerless," in *The Power of the Powerless*, edited by John Kean (London: Hutchinson, 1985), p. 27f.

set the tone of social life in the USSR. These generations have a feeling (almost religious) of the "sacred obligation" toward the compatriots who fell in the war.

7

Democratic Socialism or "Perestroika" of Statism?

FEASIBLE AND VIABLE SOCIALISM

For highly deterministic Marxists, socialism constitutes a historical necessity: the collapse of capitalism and the victory of socialism follow "iron laws." But, for humanistic Marxists, socialism is only a historical possibility and tendency, and its realization crucially depends on peoples' value choices and their collective actions. For both, however, socialism is so inseparable from communism that it constitutes just one of communism's phases.

We today are interested most of all in a socialism with realistic and foreseeable prospects. Therefore, "feasible"[1] and "viable"[2] socialism should be conceptualized apart from the utopia of a classless and stateless society.

---

1. Alec Nove, *Economics of Feasible Socialism* (London: George Allen & Unwin, 1983).

2. Wlodzimierz Brus, "Socialism—Feasible and Viable?" *New Left Review*, no. 153 (September-October 1985):43-62. I was largely stimulated to radicalize my opinions in this respect by reading Nove and Brus.

We find the most promising starting basis for socialism in highly developed and democratic capitalism, because of its economic and political level and because of the structural separation of civil society from the state. Democratic socialists claim that this separation should be kept under socialism as well. Marx would certainly not agree.

Without the existence of a strong civil society, which would inevitably also have important bourgeois components, socialism right now looks very unlikely and unattainable, and one of the reasons is that it must be organically integrated in the world economy and world market.

It is high time to draw radical conclusions from the fact that there is not a single example of a *complete break* with bourgeois political, economic, and cultural tradition that ended well in the opinion of democratic socialists. The only feasible and viable socialism for today's world and tomorrow's is *socialism with a civil and bourgeois face*. This is not a "transition period" between capitalism and communism, or a "lower phase" of communism, but a new social formation with a fairly mixed type of ownership, economy, and entire civil society.

To put it in economic terms, the dominant socialist dimension of the civil society would be the self-management enterprise based on the social and cooperative ownership of strategic[3] productive means, and that within the framework of democratic planning of economic proportions and economic growth. Its bourgeois[4] dimension would be private ownership of nonstra-

---

3. Of course, a democratically adopted criterion for differentiating between the strategic and nonstrategic, changes with the advance of science and technology. What is of strategic importance in industrial society loses that attribute in post-industrial society, because access to information assumes a decisive role. The "hammer and sickle" have long since become obsolete as symbols of socialism.

4. This should be admitted openly, rather than harbor the "bad faith" that private ownership and commodity production everyday, spontaneously, and inevitably breed *socialism*. That would be just

tegic productive means; market competition; and production for profit by private, cooperative, and social enterprises. Democratic socialists in Yugoslavia and in the world rightly call this kind of economy a market-planned economy.[5]

To be sure, in order to talk about socialism at all, the bourgeois effects of private ownership and market competition must be compensated for by a solidaristic conception of justice. For instance, by guaranteeing every individual a living standard and social services worthy of them, irrespective of ownership or the person's success in the market, and even irrespective of whether he wants to work or not.

VIABLE STATISM

So far we have been talking about feasible and viable socialism. However, countries that call themselves socialist are, in fact, statist, because in them one group holds a structural monopoly of control over the state and through it over the means of production.[6] We are already witness that statism is *feasible*, but whether it will remain *viable* depends on the possibilities for radical self-reformation.

---

as bad (if inverse) faith as that whose victims were the Bolsheviks, led by Lenin, when they feared that such ownership and production, even if on a small scale, would inevitably produce *capitalism*.

5. I myself have long advocated such an economy. See, for instance, my book *Between Ideals and Reality* (Oxford: Oxford University Press, 1973) (translation from the Serbo-Croat original published 1969), especially p. 130ff.

6. The statist principle of distribution has nothing to do with labor, ownership, or market. I would formulate it as follows: *From everyone according to his ability to be obedient—to everyone according to his place in the statist hierarchy and importance to it.* The fact that officials depict statist distribution and privileges as the result of labor, merit, and responsibility is another matter.

We could call such reformed statism—*statism with a civil and bourgeois face.* Admittedly, a *comprehensive* civil society calls for a *completely legal and pluralistic state.* However, by definition statism in self-reformation could not go that far. That would be revolution, not reform.

Those liberals in statism who do not take into account this major difference, are themselves the victims of the utopian way of thought for which they constantly accuse Marxists. It is irresponsible to the people not to say openly that there are realistic chances for the liberalization of statism, whereas there are no outlooks, at least not in the foreseeable future, for revolution that would establish a truly democratic state. The "spectacular principledness"[7] of such liberals could lead to revolutionary adventurism, which would most probably result in suspension of the liberalization and even restoration of strongly repressive statism. The history of statism provides us with instructive examples in this respect.

True, even liberals who want to be political realists and engage only in liberalizing statism,[8] are not without their problems. It is not at all pleasant to restrict oneself to reforming a system one would basically prefer to dismantle. Furthermore, how is one to assist the liberalization of statism without being manipulated by the ruling class? In short, how can one not be paralyzed by *unhappy consciousness* (if I may use Hegel's term so freely)?

This problem is familiar to the history of Marxism and

---

7. A. Podgorecki uses this formula in an entirely different context.

8. In saying that only liberalization, not democratization, is possible *within the frameworks* of statism, I, of course, do not rule out the possibility of the democratization of individual groups, organizations, or institutions, but only the possibility of establishing a truly democratic state, since this would mean eliminating the monopoly over the state, in which case it would no longer be a statist society.

communism as well. Marx himself already had that tension of being caught between the sober realization that his program is possible only on the premises of developed capitalism and his personal impatience and desire for the proletarian revolution to break out as soon as possible and anywhere. His correspondence with Vera Zasulich on the outlook for skipping over capitalist development in Russia shows that he never managed to overcome this ambiguity. The Bolsheviks believed, however, that the concept of the "weakest link" and "permanent revolution" would help them achieve this end.

In order to somehow reconcile post-revolutionary development with their disappointment in it, many Marxists shifted their preoccupations from a higher to a lower (socialist) stadium of communism. Stalinist reality became so incongruent with Marxist images of socialism, that it had to be resolutely opposed in the name of *socialism with a human face,* but without answering the question: What kind of socialism is it if it has an inhuman face?

That is why I recently suggested that we distinguish between *statism with a human* and *statism with an inhuman face.*[9] The first step in concretizing the idea of statism with a human face would be the premise of statism with a civil and bourgeois face, or simply liberalized statism.

But, let us return to the phenomenon of unhappy consciousness. The position of democratic socialists in statism is similar to that of Marxists in pre-revolutionary Russia: they wanted socialist revolution, but theory called for patiently preparing just bourgeois-democratic revolution.

How are democratic socialists to work for the liberalization of the statist monopoly, when they would prefer to eliminate

---

9. Present-day Slovenia is a good example of the evolution toward statism with a human face, whereas we can still characterize some other Yugoslav republics or provinces as statism with an inhuman face.

it totally? How are they to take part in the liberal enlightenment of the statist class when they are radically opposed to its rule? If this is not unhappy consciousness, it is certainly not happy consciousness.

Like liberals, when democratic socialists advocate reforming statism they must be aware that the ruling class will try to use them toward its own ends. I do not know where an Andrei Sakharov would see himself in today's political spectrum, but I do know that he is adept at avoiding this danger. It is both interesting and surprising how talented this great natural scientist is in the political arena. Take, for instance, his statement at the International Forum in Moscow in February 1987: "We need the free flow of information; the unconditional and complete release of prisoners of conscience; the freedom of travel; to choose one's country and place of residence; effective control by the people over the formulation of domestic and foreign policy."[10] Sakharov knows that he has taken a great moral and political risk in openly supporting Gorbachev, therefore he continues to publicly support far more outstanding causes than Gorbachev, thereby reducing the risk. We also see from the last sentence quoted that Sakharov calls for democracy, without at the same time calling for the institutionalization of political pluralism, because he knows that this is impossible without revolution.

I do not see how the theory of totalitarianism could allow or explain the possibility of liberalizing statism, not to mention statism with a human face as the culmination of that liberalization. Admittedly, the spectrum of statism can range form the despotic to that with a human face, because in both cases there is a structural monopoly of control over the state and the means of production. History had already shown, however, that this control need not be total. It was not total before the victory of Stalinism (even in the period of "war communism,"

---

10. From *TIME* magazine (March 16, 1987).

not to mention the new economic policy, and it has not been so in the post-Stalinist period. Surely we are not going to say that the *totalitarian essence* of the system emerged ten years after its *existence!*

The conceptual strategy that embraces so many different phenomena under one and the same *extreme* designation is curious to say the least. What about the huge differences between extreme totalitarianism under Stalin, and the incomparably more moderate version under Gorbachev, or the extremely moderate version in Hungary and Poland, and the pseudo-totalitarianism in present-day Slovenia in Yugoslavia? The very absurdity of asking how total is totalitarianism in the concrete case proves that this rigid categorical approach is not fit to cover the seventy years of existence and change that comprise communist statism, and it is even less fit to foresee its future changes. This approach suffers from the same type of shortcomings as statist "scientific communism," according to which the difference between democratic and dictatorial capitalism is merely one of form, not substance.

Surely the experience of people living under statism counts for something when assessing the degree to which its changes are substantive or negligible. I do not believe that there are many people who, having tasted Stalin's tyranny and who now live under Gorbachev's "perestroika," would agree that the "totalitarian essence" of the system has remained unchanged. Dare a theory be so arrogant and insensitive as to dismiss disdainfully the difference between the life and death of millions of people?

I am convinced that the distance between Stalinist and liberalized statism is greater from the *human standpoint* (and what other standpoint should we take!) than the distance between liberalized statism and democratic capitalism. If this, too (apart from introducing the market and private ownership in statism, and planning and state ownership in capitalism), is borne in mind when talking about a certain convergence of the two systems, then I have no objections.

The theory of statism (the outline of which I first published in 1967) cannot be rounded off until we define the relationship between the statist class and the Communist party. The history of the transformation of the revolutionary Communist party into the *main transmission belt of the ruling class* has yet to be studied.

As is known, the self-understanding of communist parties in power includes treating other organizations and institutions as their own transmission belts. However, this entire concept of transmission belts functions as the ideology of the statist class as long as it conceals the fact that the Communist party is itself in the position of a transmission belt. Even critics help to preserve this ideological picture if they fail to deepen their analysis with this insight into the relationship between the Communist party and the ruling class.

This is also done by critics who concentrate on the *one-party* character of the system. There is obviously just one party here, but it is more than naive to believe that ordinary people, who constitute the largest segment of its membership, belong to the ruling group. They are hardly more powerful than nonmembers.

It is also not good to stop at the premise of the party-state and state-party, because, in the final analysis, what we are dealing with is the *class state* and the *state class*. Admittedly, because of its predominantly political character, the statist class has a need to organize itself as a party and to act through it.

Neither is this a question of the dictatorship of the Communist party (in the guise of the "dictatorship of the proletariat"), but of the dictatorship of the statist class, which operates implicitly as the vanguard of both the Communist party and the working class. Finally, "democratic centralism" does not primarily serve party purposes but rather constitutes the principle of organizing, mobilizng, disciplining, and controlling a part of the population by the statist class. The

connection with the people is not small if "democratic centralism" is taken to embrace, let us say, ten percent of the
population. Moreover, through that part of the population and
other transmission belts the statist class also establishes ties
with the rest of the people. It is no wonder that, because of
such ties (and the system of the redistributive economy), people
find it hard to identify the ruling group as the ruling class.

STRUCTURAL POSSIBILITIES FOR
THE LIBERALIZATION OF STATISM

"Secondary," "hidden," "implicit," "nonlegal," "nonlegitimized,"
"nonformal" spheres, activities, and ideas are playing an expanding role in ideology, economics, politics, culture, and morals in
statism. Some of them help to make society and the state function,
although the formal system poses an impediment to them.

The ruling class occasionally tolerates such phenomena and
tries to reduce or even eliminate them by means of organized
campaigns and measures. If it cannot completely reject them
as foreign bodies, then the statist ideology tries to suppress
or conceal them as much as possible. This merely leads to the
increasing surplus of illusory ideology and failed investments
in the ideological-propaganda machine.

Practical steps are often counterproductive as well. When
statists decide, for instance, to allow private enterprise in crafts,
trade, tourism and other services, they usually try to minimize
it to, as they put it, prevent private enrichment. But, since
private businessmen are few in number, they get rich quickly.
This, in turn, leads to a revolt within the statist apparatus
and among ordinary people, which revives the campaign against
private businessmen. And so the cycle continues.

Despite the steps undertaken, "secondary," "hidden," "implicit," "nonformal," "nonlegal," and "nonlegitimized" spheres
are making increasing headway. For a flexible ruling class the

very existence of such spheres would be enough of a sign that systemic reforms are needed.

What would have happened to capitalism in the thirties had the bourgeoisie been unable to reconcile itself to the need of state interventionism? The statist class is now at a historical turning point: in order to preserve *selective-strategic control* over the state and the means of production, it will ultimately have to sacrifice *total, super-centralized, and detailed control.* Only the first is necessary to preserve the identity of the statist system. Has not Yugoslavia already demonstrated that considerable statist decentralization is possible?

It is true, as G. Markus says, that this system has so far tried to maximize the *amount* of economic products and the *scope* of control over societal life. But, I do not believe that the statist class will be unable to replace *extensive* with *intensive* economic development, and universal, super-centralized, and detailed control with a selective-strategic variety. It is possible for this class to preserve a "leading role" by reducing its scope of control over society.

What would an examination about "history and class consciousness" look like with the statist class as the main topic? Can we talk about the need for introducing into statists "class consciousness from without" regarding the need for systemic reforms? Who will open the statist class's eyes, and how, regarding its own objective and long-term interests?

I say that the *liberalization of statism* would be in the objective interest of a good part of the ruling class. Only *democratic socialism* would spell the end of statist rule. The chance for such socialism in statist countries will, however, remain more or less illusory until the system is liberalized first. The Leftist intelligentsia will alternate between resignation and excessive expectations until it realizes this point.

If we abstract external control and intervention, Czechoslovakia's experience in 1968 indicates the possibility of a certain kind of *permanent reform* in a developed statist country

with a democratic tradition, where mass pressure for the un-
interrupted liberalization of statism turns into a mass movement
for democratic socialism.

But, there is no theory of the liberalization of statism, let
alone theory of the transition from liberal statism to democratic
socialism. It should be added that *democratic* capitalism, with
its comparative advantages, will continue to pressure liberal
statism to change in the direction of a pluralistic state.

As for the *ideologues* of the liberal-statist orientation, it
should be said that today it is virtually impossible to defend
openly the right of any social group to monopoly control (be
it only selection-strategic) over the state and means of production. That is why I do not believe that a liberal-statist *theory*
will be built which could, in terms of its philosophical and
other intellectual qualities, compare with the greats of bourgeois
liberalism.

Both an *inter*-class and *intra*-class struggle is being waged
over statist reforms. Since it emerged as a revolutionary
thought, Marxism usually neglects intra-class and overesti-
mates inter-class conflicts. Whoever wants to examine the possibility of reforming statism must take into account conflicting
interests between various parts of the ruling class: politocratic,
bureaucratic, technocratic, military, police, propagandist, and
others. The transition from extensive to intensive social control
and production is bound to lead to major shifts in the configuration of the statist class. The technocracy, for instance,
which until now has had relatively little power, will greatly
gain in number and importance, at the expense of the politocracy
and bureaucracy. Furthermore, the huge apparatus in charge
of "agitation and propaganda" will also have to undergo serious
changes. Some conservatives will resist the reforms; they will
be required to ideologically justify and propagate the reforms.
Others might adapt, but they will be hit by measures to reduce
surplus employment in that apparatus, because of the shift
from extensive to intensive ideological production and control.

A good deal has been written about the role of ideology in this type of system, but, as is indicative in such cases, there is still no study about "ideological workers." The topic of *ideology as a profession* has yet to be raised.

Since the source of the ruling class's power lies in politics, it is only natural that clashes over reforms should assume a directly political character. Admittedly, a large number of this class's members will, because of historical experience with reforms, wait and see which side has the better chance of winning.

Differences in stand regarding reform should also be examined in other classes. Manual labor will have less and less of a future as the scientific-technological revolution advances. Combined with market-oriented reform, this revolution will bring into question the survival of any number of economic enterprises, and this segment of workers will also turn against structural changes.

There is the danger of latent anti-reform alliances turning into actual ones. That is why the reformers must elaborate their plans and pursue an effective social policy, so as to prevent the most conservative segments of the statist class from linking up with the worst off segments of the workers, peasantry, and intelligentsia. *Market reform planning* is another indicator that it is untenable simply to pit the market against planning.

The introduction of limited legality seems to constitute the common denominator of all attempts to liberalize statism. In the USSR this began after Stalin's death: The leadership introduced elementary legality in order to prevent a repetition of despotic willfulness and self-liquidation of the party-state hierarchy. Stalin's suspiciousness was gradually replaced by "trust in cadres," which was to culminate with Brezhnev's conservatism.

A certain amount of legality is necessary both in terms of the self-limitation, relationalization, and division of power in the statist class, and in terms of its legitimation. To be sure, one should not expect statism to establish a legal state in the full sense of the word. No matter how liberalized, that state

will retain the basic characteristic of duality ("the dual state"),
in which one group's monopoly of control over the state and
the means of production has a meta-legal status. But, when
legal reform begins to act, after a while it acquires a dynamic
of its own and opens up the possibility of pressure from below.
More and more professional lawyers, intellectuals, and citizens-
at-large will take advantage of this opportunity to reduce the
power of the meta-legal state.

It is extremely important that domestic and world markets
are calling for legal security. A market-oriented statist economy
cannot be developed as long as there is an arbitrary and even
nihilistic attitude—politically, economically, legally, morally,
culturally—toward forms, rules, procedures, and contracts.
That is why Milovan Djilas called, while still in power, for
the rehabilitation of "form" in Marxism and socialism.

The international "capitalist encirclement" of statism is
strong. This over-determination pushes it to reduce political
repression and liberalize penal law. Here I'm not thinking so
much of the fact that the world market is dominated by *democratic* capitalism, which exerts pressure for respecting a minimal group human and civil rights. I am thinking instead of
the economic damage sustained by statist countries that
persecute their citizens. Protesting against "interference in one's
internal affairs" is of no use here. At issue is an assessment
of the political risk entailed in investing, crediting, and generally
doing business in such countries. Even if they give all sorts
of legal guarantees, remove ambiguities in their economic
legislation, eliminate the self-will of local power-holders, and
curb inflation, statist countries will find it difficult to attract
foreign capital and to persuade their own citizens to start private
businesses. And this will continue until such countries desist
from anti-liberal political-ideological campaigns and from
persecuting their own citizens.

In evaluating political security, business people rightly
proceed from coherent wholes. In the language of historical

examples, this means that the "new economic policy" does not have good prospects as long as there is "war communism" in other spheres of societal life. What else can foreigners think about the political stability of a country if its leaders use the military or even just threaten to use it in order to protect themselves against their own people? When they persecute their own citizens—even the most prominent ones—what guarantee can be given foreign investors that their capital, along with that of domestic private business, will not be nationalized again? It appears that real competition will develop among statist countries in attracting foreign capital and stimulating domestic private investment.

SELF-MANAGEMENT UNDER STATISM

I would like briefly to raise another question related to the liberalization of statism; that is the question of feasible and viable self-management. Yugoslavia's experience in this respect is instructive.

First, there is the relationship between "civil society" and self-management. A realistic approach to self-management under existing statism (rather than some ideal construction) means discussing possibilities for constituting the civil sphere independent of the state. If one sets out from the utopia of the withering away of the state, as in Yugoslavia, then the result will probably be self-management as a utopia.

Since there is so much talk about the need to introduce a "real economy" in Yugoslavia, one should openly confront the dubious elements and the fictions not only in economics, but also in politics, ideology and—yes—self-management. Isn't self-management fictive if it has so many debts and losses? They are consequences, among other things, of the fact that self-management has remained, to this very day, a transmission belt of the statist class. Moreover, there is the question of what kind of self-management can exist without the right to *self-organize* and the possibility of doing so. What kind of self-management could exist if the state, which is already monopolized, prescribes—down to the last detail—how so-called self-managers should organize themselves? Now it appears that self-management will be introduced in the Soviet Union and in other statist countries in the same way—from above.

Second, we should have pluralism of self-management and the characteristic conflicts that arise within it. The various forms of self-management should be adjusted to the types of ownership and to the existence or nonexistence of market competition. Self-management would then vary from participation in management (as in a large electric power plant or other public utility) to the unlimited self-management that exists under conditions of full market competition (as in footwear factories). The law would prescribe only general principles, leaving it to the respective enterprises to choose how they will regulate their own management within the specified legal framework. During registration, economic courts would judge whether enterprises have organized themselves in accordance with these principles. If an enterprise has a monopoly position, then the majority in its managing body would be (paid) representatives of the banks, enterprises, and state bodies that have invested in that enterprise. Its employees would only participate in management.

A fundamental feature of realistic self-management is the conflict between group and societal interests. If the interests of employees are represented in management, then there should be some *counterweight*, preferably in the form of participation by representatives of other groups and the state in management. The state's role would also be made manifest in the form of laws and other regulations. If self-management does not proceed from this presumption of conflict, the inevitable result will be its own negative dialectic: the functions of management will be constantly suspended by statist intervention initiated in the name of social interest. As an example of these conflicting interests, one need only observe that in Yugoslavia consumers have still not started—spontaneously or independently—to organize themselves for protection against the self-will of monopoly enterprises.

Third, it would be instructive to ask what the relationship should be between self-management and management bodies? In my opinion, it should be defined in a manner analogous to the relationship between legislative and executive government. Once elected, the board of managers (the executive) runs its affairs without interference from, but under the control of, the *council of employees* (the legislature). Unless such autonomy is guaranteed to management bodies the result will inevitably be dilettante-like decisions, which cover up irresponsibility and incompetence by citing collective self-management decisions, and, in the final analysis, mediocrity is both prevalent and pervasive. In Yugoslavia, people talk about the danger of every kind of majority rule except that of mediocrity. Indeed, the danger exists of the *envious communism of self-managers*.

It is for this reason that the critical notion of *prolet-production*—analogous to the "prolet-culture"—needs to be introduced. Prolet-production and its corresponding notion of the "immediate producer" (as the sole creator of surplus value) poses a major obstacle to the development of modern production and rationally organized self-management. In the phase of extensive economic development, based on a cheap workforce, the statist group accentuates in its *workers' ideology* physical rather than intellectual labor as the main if not exclusive source of surplus product. With this kind of approach there is no question of effectively joining in the international division of labor, especially not now when know-how, information, and new ideas are becoming the main productive force.

MARXISM AND THE POLITICAL PROGRAM OF REFORMS

In some statist countries, most notably in Poland, the largest
segment of the population does not want to hear any Marxist
language, let alone official Marxism-Leninism. Workers in
Poland do not readily accept the explanations of Marxists who
refer to Solidarity as "the movement for the self-emancipation
of the working class." True, the real character of groups and
movements ought to be determined on the basis of their attitudes
toward concrete social problems and not according to their stand
on any abstract formulae. If Solidarity is approached in this
manner it will not be seen as incompatible with humanistic
Marxism. However, the question of how Marxism could survive
in an environment in which even the apparatus of mendacious
consciousness seems to use the same language cannot easily
be dismissed. It is no small difficulty for a theory to be sustained
when people, already at the verbal level, reject it out of hand.

When considering this problem, I believe a difference should
be drawn in classical Marxism between the principles of radical
humanism and the idea of communist social organization. These
familiar principles include: praxis, de-alienation, de-reification,
meeting authentic human needs, the freedom of each and every
individual as the condition for the freedom of all, and others.
Marx linked their realization with classless and stateless social
organization, where private property and a commodity-mone-
tary economy were to be abolished, and the distribution of
the social product was, in the first phase, to be carried out
in accordance with the work invested, and later in accordance
with needs. Life has proven Marx's idea of a communist society[11]
to be in part irrelevant to the prospect of feasible and viable
socialism, and in part incompatible with it. This even applies

---

11. Even when we abstract his epochal failure with the idea of
the "dictatorship of the proletariat" in the "transition period" toward
such a society.

to the most developed countries in the foreseeable future. This, of course, is not to say that Marx's humanistic principles need have the same fate. Provided we separate them from communist utopia and understand them as ultimate regulative and critical, not constitutive and operative principles, they can, with numerous necessary mediations, be useful in assessing existing societies and projects of democratic socialism.

Admittedly, criticism of existing communism in the name of Marx's communism took on a subversive character in the fifties and sixties in some statist countries (e.g., Hungary, Poland, and Czechoslovakia), because it successfully questioned the system's Marxist legitimacy and challenged the ideological monopoly of the ruling class. The bitter reaction of officials and the persecution of critics showed that pointing to the gap[12] between statist "Marxism-Leninism" and Marx's communism did indeed irritate some sore spots. Such criticism will continue to be useful, politically as well as intellectually, as along as there are people who believe in the relevance of communist utopia. After all, in some statist countries such criticism has not had real access to the public and, hence, could not be tested, let alone exhausted. Yet, one thing is beyond doubt: it is incomparably more important to the people whether Marx's idea of communism can be the starting point for elaborating political programs of social change than whether ruling communism is in accordance with it.

In the approaches that Agnes Heller and I take to the above problem, interesting differences have manifested themselves. It would be useful to quote what she has to say in this respect:

---

12. If such enlightenment is superfluous for educated and informed people, that is not to say it is deplacé for the uneducated and uninformed, especially for the youth who are subjected to everyday "ideological education."

Question:

> During a debate in 1979 in Köln between Rudi Dutschke, Boris
> Weil, Plushch and other oppositionists from Eastern Europe there
> came to the fore something which has become obvious in the
> contemporary conflicts in Poland. During the days of internation-
> alism in Tübingen, Stojanović summarized this as follows:
> according to his experiences with Polish oppositionists in Poland
> the entire Marxist language has been discredited. It is no longer
> possible to lead any kind of discussion using this language. The
> left in the East is searching for a new language.

Heller's answer:

> The opposition does not want to speak in the language of the powers
> that be. . . . What can be done, I believe, is to reclaim this language
> from the powers that be. . . . We can give new meaning to these
> words. . . . We can always strip the powers that be and the gov-
> ernment naked by confronting them with their own language.[13]

Heller obviously believes that we can approach the ruling
communist ideology in an *immanently* critical way, just as Marx
approached bourgeois ideology. Let me just remind the reader
that the bourgeoisie cited freedom and equality as its achievement and that Marx—holding it to its word—tried to show
that these values had been achieved, but only at one level ("form"
and "appearance"), while at the other, deeper level ("content,"
"essence," and "reality") they are a cover for class domination.

Until approximately twenty years ago, the ruling communist ideology primarily cited the objective and historical interests of the proletariat. Since the ideologues could always shift
the realization of these interests *ad calendas graecas*, criics
were never able to deal them a final blow by citing the non-
realization of these interests.

Admittedly, Heller made her statement at a time when the

---

13. "Intellektuelle und das stalinistische 'Erbgut'," in *Inter-
nationalismus-Tage*, Tübingen, 11-13 Dez. 1981, (Dokumentation,
Tübingen, 1982).

ruling ideology had already travelled the road from "socialist realism" to "real socialism." That is why we should ask ourselves whether at least in the latter phase it is possible to "wrest" the language of Marxism from the ruling class by pointing to the gap between it and reality, and whether in this way Marxism would not become attractive to the people. Heller, who unquestionably has contributed to critical Marxism, does not seem to have noticed a change that only heightened the problem.

Symptomatically, the force of the Marxist critique of ruling communism has increasingly waned as real prospects have opened up for its liberalization: e.g., the decentralization of management, reliance on the market, a certain rehabilitation of private property and initiative, and establishing the sphere of "civil society." However, even with the greatest possible conceptual elasticity we cannot integrate such measures and changes into Marxism, and still less can we characterize them as Marxist. Unless this is understood, critics of existing communism who rely on Marx's communism are threatened with a new, now deadly danger: that of becoming conservative. Marxism is unable to *explain* the emergence and nature of statism, all it needs now is to *attack and hold back* necessary changes. Unfortunately, it does not require great effort to use Marx's critique of bourgeois political economy against such liberal measures and changes. It is not hard to guess what Marx would say about them since he called even the principle of distribution according to work bourgeois.[14]

---

14. A passing observation: It does not ensue, as Marx thought, that such distribution must be bourgeois if it is not communist. Of course, here I entirely leave aside the question of what distribution according to work concretely means, what its practical criteria are, and, finally, whether it is at all feasible. Admittedly, this question does not at all affect distribution according to success in the market, but Marx was against it. However, it has proven to be unavoidable, both for the liberalization of statism and for the project of a feasible and viable socialism.

It was only natural that Marxist *philosophers, philosophical* sociologists, and *philosophical* economists played the leading role among the critics of statism. However, it is instructive that the alarm of the ruling class became shriller as philosophical critics demonstrated an ability to climb "down" in their criticism from the most general and theoretical to the more concrete political level, and even showed an ability to engage in practical opposition activity. The ruling class came to feel that the people desired a more *realistic political program* of changes in statism. Such a program does not exclude but rather assumes serious *symptomatic interpretation* of official ideology.[15] Even differences in it which appear small can be a good indication that serious divisions exist within the ruling class. Thus, for instance, in the Soviet Union a public debate had begun before the latest congress of the Communist party regarding whether in "socialism as well nonantagonistic contradictions can become antagonistic ones." Those who sufficiently understand this ideology and its history concluded immediately that this represents the hidden debate as to the weight of the problems confronting Soviet society today, and the depth of the needed changes.

For successful symptomatic interpretation it is very important, as scholars have pointed out, to distinguish clearly between *abstract* and *operative* ideology. The former should be sought, for instance, in the "theoretical" texts of the leaders, the state constitution and the Communist party's program, whereas the latter manifests itself in laws (especially criminal law), executive orders, the party's statutes, and the like.

---

15. In this way the following difficulty may be avoided: How can we intellectually deal with the official ideology when it itself ultimately has neither intellectual purpose nor real intellectual quality? Do we not thereby accord to this ideology intellectual stature and dignity? (See F. Feher's article "Eastern Europe in the 1980s," *Telos*, no. 45, in which he criticizes this "indirect apology.")

If we want to discover, for example, whether *real* intentions and chances for statism's liberalizaton exist, we should not rely too much on innovations in abstract ideology, which are often designed to convey a favorable impression, especially abroad.

The most important question is: Where should we search for the *operative ideology par excellence?* Many scholars still suffer from naivete, since they continue to search for it in the sphere of "high" ideology. In my view, much more indicative in the operative sense is the ideology embodied, for instance, in the textbooks for elementary party schools, for officers and soldiers, for policemen, and other official purposes.[16] Only at this level can we discover, if not the real intentions of the leadership, then at least the extent of their real power to prompt all intended liberal changes.

"PERESTROIKA" IN THE SOVIET UNION

With the recent significant leadership changes in the USSR, the question of reform has again become the focus of attention. That part of the ruling class and party in the Soviet Union which desires significant reforms is searching for support in their own ideological tradition. Although no one would doubt that the Soviet Union has something to learn from the reforms in Yugoslavia, China, and Hungary, I do not believe that its leaders will want to refer publicly to primarily foreign examples.

As the American philosopher William James pointed out, nothing new can be accepted as truth unless we make it fit with the minimum of disturbance and maximum continuity into our original stock of truths. Such continuity cannot be preserved in the USSR without referring to Lenin. Fortunately,

---

16. In Yugoslavia, most of the extensive research of this ideology has been done by Nebojsa Popov.

not only "war communism" but also the "new economic policy" are dimensions of Leninism.

Generalizing on the basis of Soviet history, one might say that everywhere ruling communism manifests these two patterns of rule, development and reaction to crisis. How can one explain the predominance of "war communism" thus far?

The first point to be made is that "war communism" is the *formative* phase of communism, whereas the new economic policy is an effort to *transform* the already established "war-communist" tendency and structure. After all, before coming to power, communist parties usually spent quite some time underground, where they developed their organizaton and their mentality. Finally, these parties as a rule came to power during war and by means of war.

Much of the history of the Soviet Communist party, the original and long-standing model for all other communist parties, has been marked by war: World War I, civil war and defense against outside intervention, accelerated Stalinization through state war against the peasantry, long-standing universal state terrorism, preparations for war and bearing the main burden during World War II, state terrorism after that war, and the cold war. In a sense, Stalinism is *super-war communism.*

Chinese communism, too, was until recently a kind of *permanent war communism:* e.g., the long civil war and struggle against an occupier, the "great leap forward," and the "cultural revolution."

Before coming to power, Yugoslavia's Communist party had been subjected to repression and forced to become clandestine. Then came the National Liberation Struggle, revolution, and, after that, war-communism culminated paradoxically with defensive war preparations after the break with Stalin.

Now, the Soviet Communist party is rapidly turning from Stalinism to Leninism. In 1928, Stalin cited the Lenin of "war communism"; Gorbachev is citing the "new economic policy." It is on this contradiction in Leninism that Vladimir Voinovich based a superb little story, which I shall briefly relate here. Driving back to Moscow, the writer came upon a road that had no signpost to direct him. Guessing which direction to take, he drove along the empty road, where every so often there would be a big portrait of Lenin and the inscription "You are on the right road, comrade!" After driving like this for a long time he learned from a passer-by that Moscow was in the opposite direction. He turned around and drove back. As before, the monotony was broken from time to time by the same portrait and the same inscription.[17]

Today, of course, it is no use copying the "new economic policy." Incomparably more radical economic reform is needed, and this implies fundamental political reform. In other words, one has to opt for a *new policy* (NP), not merely for a "new economic policy" (NEP).

In addition to the Lenin of the NEP and *State and Revolution*, Marx the humanist and supporter of the Paris Commune should also be expected to be the main ideological support of reform in the USSR. It is too early to expect the officials to refer to socialist thinkers of non-Marxist provenance.

The other immanent ideological possibility for opening up to reform lies in the so-called "scientific character" of Marxism-Leninism and its advocacy of the scientific-technological revolution. As early as Brezhnev's rule, stagnation was transformed into "competition crisis." The USSR's general lag behind the West continues to grow more serious, and is the main reason for "perestroika." Because of its official anti-intellectual policy, the USSR has suffered a severe "brain drain." As a great patriot, Gorbachev wants to stop this sapping of intellectual power by widening the scope freedom for members of the creative élites. He will also provide the possibility for traveling and even temporarily working abroad. However, such freedom of

---

17. "Taking Lenin's Direction," *The New York Times* (October 19, 1985).

movement should not be expected for ordinary people in the near future.

In pointing to the "competition crisis," I do not wish to underestimate the pressure of circumstances and of the Soviet public for finally embarking down the road of reform. Fewer and fewer people can expect improvement through the hierarchical promotions typical of extensive development. Now the accent should be primarily on improving one's own situation in the existing position.

Ever since the recent changes occurred in the Soviet leadership, Western experts have been in search of the "really existing" Gorbachev. Leaders can certainly make a significant impact, especially if they, like Gorbachev, have much greater reformist potential than the ruling class, the party, and the system. Precisely this disproportion, however, represents the main obstacle to Gorbachev's intentions. Of course, I do not believe that Gorbachev has a fully defined plan of systemic reforms that, for tactical purposes, he does not yet want to reveal. He himself is still searching for the "really existing Gorbachev."

At the latest party congress, Gorbachev sharply criticized those who would "like to improve things *without changing anything*." After less than one year of experience as secretary general, Gorbachev found out how much resistance to reform he actually faces: "Unfortunately, there is a widespread view that any change in the economic mechanism is to be regarded as virtually a retreat from the principles of socialism." For the time being, Gorbachev appears to have support from the minority in the hierarchy, but that support is strategically situated. He has a good deal of personal power, even though in the collective leadership he is only *primus inter pares*.

His is the guiding hand in the work of the Central Committee Secretariat, which has an important say in preparing the sessions of the Politburo and Central Committee and in running personnel policy. Gorbachev knows that in this type of system "cadres are of decisive importance" (Stalin). That is why he does not engage in political adventure, but rather makes sure that he has advance support from the Politburo and corresponding personnel changes for every move. He alone in the leadership can talk on his own initiative with any individual or group, without the risk of being accused of factionalism. This is a tremendous edge for any general secretary in assessing the relationship of forces. Finally, Gorbachev's power is rooted as well in the huge symbolic prestige of his office. In contrast to Khrushchev, Gorbachev is not saddled with Stalinist state crimes. For this reason he can call for "glasnost" with regard to that part of history and seek legitimacy by making such a call. In dictatorial politics, be it autocratic or oligarchic, the change of generation is of enormous importance.

The attitude toward the ideological story of the power-holders is a convenient opportunity to remember all that it can conceal. It is well known that there can be a major difference between the real dangers to power-holders, on the one hand, and the way those dangers are perceived, on the other hand. Yet, only one part of this surplus power, control, and repression can be explained by their excessive assessment of the dangers. The rest is usually due to the arrogance and even the maliciousness of power-holders: the ideological story then shrouds their abuses, misdeeds, and crimes in deep darkness. How can one expect evil-doers to "quietly and honorably" retire from power? Out of fear of "glasnost" they will try *at all costs* to hold their monopoly on total, supercentralized, and detailed control over the state and its history.

Gorbachev's life as functionary can cast new light on the topic of *politics as profession, the ethics of responsibility and homo duplex.* In order to survive and advance within the conservative political hierarchy, he had to conceal his true self. Moreover, at times he accepted less promising posts so as to not sully his hands. Looking back, his was a great and successful bet with history. To be sure, there was also a good deal of luck involved.

One should, however, guard against unmeasured judgments. Let me give two examples. One is the recent statement by the French Communist party "dissident" Pierre Juquin that this is a "revolution within a revolution."[18] However, there is no question of revolution, but merely reform.

Maczasz Siresz, secretary of the ruling Hungarian party's Central Committee, claims that "perestroika" refutes Berlinguer's assessment that the stimulus given by the October Revolution has been definitively exhausted.[19] This is an ideological appraisal. Surely we cannot explain all the positive things that have happened, are happening, or will happen in the USSR as being due to the original revolutionary impulse. "Perestroika" actually proves Berlinguer right, for it is an effort to pull out of the dead-end reached because that historical impetus stopped and a new one was nonexistent.

How can *radical changes* be made without endangering the selective-strategic control over the state and society? Such deep reforms will require the rehabilitation of private property and private initiative in agriculture, crafts, trades, and in other services. It will be necessary to permit market competition between state enterprises as well. These measures, however, would in effect necessitate the abandonment of some of the central dogmas of Marxism-Leninism. Even the mere suggestion of the need to introduce market competition in "socialism" causes severe protests. How much larger are they going to become, once it is necessary to live with the practical consequences of the market: bankruptcy of entire enterprises, unemployment, permanent restructuring of the economy?

Often the question is asked in bewilderment: How is it that "socialism" is incapable of fully implementing market reforms, whereas capitalism continually functions, due, among others things to market competition? The main difference lies

---

18. *Spiegel,* no. 7 (1987).
19. *Borba,* Belgrade (April 9-10, 1987).

in the fact that, ever since its *beginning*, capitalism has been organically linked with the market, whereas statism has accumulated unprofitable enterprises. If market criteria were to be introduced fully, a good part of the national economy would go bankrupt. The mere thought of the social dangers inherent in such change startles even the most liberal statists.

A great many ordinary citizens will also bitterly resist such reforms. *Universal state paternalism*, as has already been pointed out by many scholars, is deeply ingrained in the population's consciousness. The collectivist political, economic, and moral culture will have to be changed. It is well known that even emigrants form the Soviet Union, although they represent a very selective group, have great difficulties weaning themselves from the social-economic security afforded by statism in their efforts to adjust to the risks of market competition in the West.

If even common citizens do not easily surrender their "privilege" of having some material security through little labor, we can easily imagine the extent of resistance of the ruling class to any attacks on its privileges. I need not refer to its most basic class privilege, i.e., monopoly control over the state and the means of production, that could only be eliminated through an eventual revolution.

Due to the leveling of Bolshevik egalitarianism in the past and the remnants that still exist in the current ideology, statists attempt to hide numerous sources of income, services, and favors that they enjoy in addition to their nominal salaries. If all of these were to be translated into money and included in the nominal salaries, enormous differences between their incomes and those of common citizens would become evident.

Prior to the latest Communist party Congress, there emerged public criticism of some statist privileges. Immediately, however, the privileged few began to justify themselves: these are allegedly not privileges, but deserved compensations for hard work and responsibility; officials do have much more important business than standing in line in front of stores, restaurants, counters, doctors' offices, travel bureaus, and the like. Thus, instead of being blamed for the fact that "developed socialism" is still choked by shortages of elementary goods and services, the ruling class refers exactly to these shortages as the justification of its privileges.

In the modern age, the political culture fund owes most to Western tradition and languages, especially English, French, and German. Certain Soviet political notions are also in circulation but they, as a rule, have negative connotation, such as "prolet-cult," "socialist realism," and "real socialism." Thanks to Gorbachev the world is now learning to use such Russian words as "perestroika" and "glasnost" to denote positive phenomena. With his talent and skills in public appearances, Gorbachev has, within a short period of time, "caught up with and overtaken" many Western politicians. He does not suffer from an inferiority complex like some of his predecessors, so he has no need to resort to ideological rigidity or arrogance.

The West often asks itself, complacently, whether it should "help" Gorbachev. In fact, the only relevant and realistic question is whether it is *in the interest of the democratic and peace-minded part of the West* not to impede the new Soviet leader. Surely the most important thing is that finally there is a Soviet leader who defines the security of the superpowers and their blocs in terms of "interdependence," thus renouncing the "class struggle in the international arena." There is sometimes fear in the West of the Soviet Union getting stronger under Gorbachev. But, a stronger rival need not mean a more dangerous one, especially if the rival is aware of this mutuality.

Needless to say, such changes suit neither war communism and fundamentalist Stalinism nor war and fundamentalist capitalism. It is naive to believe that the West can force the Soviet Union to "arm itself to death." If a democratic state like Great Britain can ignore one third of its population when redistributing the national income, why should not an authoritarian country like the USSR, in the name of further armament, ignore the living standard of at least half of its population? The best policy toward Gorbachev's "glasnost" is "glasnost": one should be open-eyed and give full publicity to everything, including the relations between the USSR and Eastern Europe.

Here, too, Gorbachev has announced "perestroika." The formula of "mutual responsibility" is to replace Brezhnev's doctrine of "limited sovereignty." This, of course, should not be taken literally, because there are no outlooks for revolution, just liberalization. There is a chance of gradually replacing total, super-centralized, and detailed control (of a primarily direct and military character) with selective-strategic superpower control. However, it is unrealistic to hope that within the foreseeable future the Soviet Communist party will learn to separate its international interest from the "leading role" of the communist parties in Eastern Europe. No Soviet leader could remain in power if he allowed this role to be brought into question.

The USSR is entering a long period of liberal reforms of statism, and not of further development of "developed socialism." Big oscillations, contradictions, resistance, and conflicts will inevitably accompany such reforms.

The uneven level of development of class consciousness of the statist classes constitutes the most significant international limitation to the liberalization of statism. The ruling class at the center of the international statist system had been exceptionally rigid prior to Gorbachev and, in any case, more rigid than the ruling classes in some countries at the periphery. Whenever liberal reforms in Warsaw Pact countries failed, the explanation could be found in the *international statist encirclement*. There are now indications that this limitation might work in the opposite direction as well: from the periphery to the center. Since Yugoslavia finds itself outside this encirclement, it most clearly reveals the *internal* possibilities and the *internal* hurdles for the further liberalization of statism. The same holds true for the eventual transformation of this liberalization into an effective mass movement for democratic socialism. But this is the topic of my next book.